Explosive Instabilities in Mechanics

Springer-Verlag Berlin Heidelberg GmbH

Brian Straughan

Explosive Instabilities in Mechanics

With 12 Figures

 Springer

Professor Brian Straughan

Department of Mathematics
University of Glasgow
Glasgow G12 8QW
Scotland
e-mail: bs@maths.gla.ac.uk

ISBN 978-3-642-63740-7

Library of Congress Cataloging-in-Publication Data. Straughan, B. (Brian) Explosive instabilities in
mechanics / Brian Straughan. p. cm. Includes bibliographical references and index.
ISBN 978-3-642-63740-7   ISBN 978-3-642-58807-5 (eBook)
DOI 10.1007/978-3-642-58807-5
1. Mechanics, Applied–Mathematics.   2. Blow up (Mathematics)   I. Title.
TA350.S763 1998 532'.051'01515353–dc21 97-45045

© Springer-Verlag Berlin Heidelberg 1998
Originally published by Springer-Verlag Berlin Heidelberg New York in 1998

Typesetting: Camera-ready copy from the author using a Springer TeX macro package
Cover design: E. Kirchner, Heidelberg

SPIN 10643402   55/3144 – 5 4 3 2 1 0 – Printed on acid-free paper

Preface

The subject of blow-up in a finite time, or at least very rapid growth, of a solution to a partial differential equation has been an area of intense research activity in mathematics. Some of the early techniques and results were discussed in the monograph by Payne (1975) and in my earlier monograph, Straughan (1982). Relatively recent accounts of blow-up work in partial differential equations may be found in the review by Levine (1990) and in the book by Samarskii *et al.* (1994). It is becoming increasingly clear that very rapid instabilities and, indeed, finite time blow-up are being witnessed also in problems in applied mathematics and mechanics. Also in vogue in the mathematical literature are studies of blow-up in systems of partial differential equations, partial differential equations with non-linear convection terms, and systems of partial differential equations which contain convection terms. Such equations are often derived from models of mundane situations in real life. This book is an account of these topics in a selection of areas of applied mathematics which either I have worked in or I find particularly interesting and deem relevant to be included in such an exposition. I believe the results given in Chap. 2 and Sects. 4.2.3 and 4.2.4 are new.

This research was partly supported by a Max Planck Forschungspreis from the Alexander von Humboldt Foundation and the Max Planck Institute. It was also partly supported by an award from the Carnegie Trust for the Universities of Scotland and by a Royal Society of London research grant. These sources of support are gratefully acknowledged.

Many of the graphs in this book were produced with the aid of a graphics routine written by Professor David Walker of the Computing Science Department, University of Wales in Cardiff. I am very grateful to Professor Walker for allowing me to use his routines. I am indebted to Dr. John Nimmo of the Mathematics Department of Glasgow University for substantial help in producing some of the figures. I am very grateful to Mr. Frank Holzwarth at Springer-Verlag for assistance with the Springer production routines and, indeed, to the Springer production team in Heidelberg for their considerable help with copy editing and production. I am also grateful to Professor Kolumban Hutter of the Technische Universität Darmstadt for useful discussions on glaciology and ice physics, and in particular for a discussion pertaining to Sect. 4.5, to Dr. Klaus Jöhnk of the University of Constance for help in initially setting up the Springer production routines, to Professor Michael Khonsari of the Department of Mechanical Engineering, Southern Illinois University, for bringing the work of Professor Bair and himself to my attention, to Dr. Sergei Merkulov of the Mathematics Department of Glasgow University for translating the title and reference to the paper by Kalantarov and Ladyzhenskaya, to Professor Keith Miller of the Mathematics Department of the University of California at Berkeley for a useful discussion concerning Sect. 2.5, and to Professor Henry Simpson of the Mathematics Department of the University of Tennessee for a helpful discussion pertaining to Sect. 8.1. Finally, I am grateful to an anonymous referee for several trenchant remarks which led to improvements in the manuscript.

Glasgow, March 1998   Brian Straughan
                                       Simson Professor of Mathematics
                                       University of Glasgow

Contents

1. Introduction

1.1 Blow-Up in Partial Differential Equations in Applied Mathematics

There has been a tremendous amount of recent activity dealing with the subject of solutions to partial differential equations which blow-up in a finite time. The mathematical theory for this is extensive and reviews may be found in Levine (1990) and Samarskii *et al.* (1994). However, finite time blow-up and other very rapid instabilities occur in situations in mechanics and other areas of applied mathematics, and studies of these phenomena have very recently been gaining momentum. The object of this book is to present an account of various instances in applied mathematics where the solution, or one of its derivatives, to a partial differential equation or to a coupled system of partial differential equations, blows up in a finite time leading to a catastrophic instability. We also deal with some situations where the solution grows very rapidly but need not cease to exist in a finite time. The problems for which we include an exposition are of importance in real life and hence justify our inclusion of them. The main emphasis here is to include recent developments in blow-up or rapid growth of solutions to practical problems which occur in some field of mechanics.

Among the exciting developments of rapid instabilities we cite the work of Butler & Farrell (1992, 1993, 1994), Farrell (1988a, 1988b, 1989, 1990), Farrell & Ioannou (1993a, 1993b, 1993c, 1994a, 1994b, 1995, 1996a, 1996b, 1996c), Gustavsson (1991), Henningson (1995), Henningson *et al.* (1993), Henningson & Reddy (1994), Hooper & Grimshaw (1995), Kreiss *et al.* (1994), Reddy & Henningson (1993), and Schmid & Henningson (1994) on shear flows, Poiseuille flows, and related flows in atmospheric dynamics. Connected to this is the work on shear flow in a compressible fluid layer by Hanifi *et al.* (1996). Also, a mathematically similar topic concerns instabilities in shear flows in a granular material, see Babic (1993), Chi-Hwa Wang *et al.* (1996), Savage (1992), and Schmid & Kytömaa (1994). Such instabilities in granular media have application in landslide technology, and in avalanche theory, see e.g. Foda (1987, 1994), Hutter (1983), Hutter *et al.* (1987), Vulliet & Hutter (1988b), and are thus very important. This exciting new development of growth of solutions in parallel flows is dealt with in Chap. 8.

Other mundane areas of very rapid solution growth occur in vortex sheets, bubbles, Rayleigh–Taylor instability, Kelvin–Helmholtz instability, as described by Baker *et al.* (1993), Caflisch & Orellana (1989), Moore (1979). In the theory of Volterra equations there has also been found singular solution behaviour, see Olmstead *et al.* (1995), Olmstead & Roberts (1994, 1996), Roberts *et al.* (1993), Roberts & Olmstead (1996). Again, these studies have practical application in solid mechanics, cf. Olmstead *et al.* (1994). A possible finite time instability which may have important environmental / climatic consequences is that pertaining to the rapid temperature increase at the base of an ice sheet, and the consequential shear instability of the ice sheet, as studied by Yuen *et al.* (1986). A very interesting development of finite time blow-up of a solution has been in the mathematical biology area of chemotaxis, see Jäger & Luckhaus (1992), Levine & Sleeman (1995). These topics and others are covered in the course of this book.

Before commencing directly with blow-up situations in mechanics we do, however, explain some of the mathematical findings which have been made in partial differential equations. We also include in Sect. 1.3 a brief review of recent developments on blow-up in the mathematical theory of partial differential equations involving systems or equations which include convection terms, these being pertinent to the applied mathematical topics which follow. Chapter 2 is an account of solution behaviour to a system of partial differential equations which contain convection terms. This is quite revealing in itself.

In the mathematical literature there has been extensive work on establishing solution structure in the neighbourhood of the blow-up time, the development of asymptotic formulae valid in the vicinity of blow-up, and work on the question of how (and in what sense) the solution may be continued beyond the blow-up time. We do not detail all of this, but refer to the papers of Bebernes & Bricher (1992), Berger & Kohn (1988), Bricher (1994), Budd *et al.* (1994), Budd & Galaktionov (1996), Deng *et al.* (1992), Etheridge (1996), Filippas & Kohn (1992), Floater (1991), Friedman & McLeod (1985), Friedman & Giga (1987), Friedman & Lacey (1988), Giga & Kohn (1985, 1987), Levine *et al.* (1989), Mueller & Weissler (1985), Smith & Bowles (1992), Stewart & Smith (1992), Straughan *et al.* (1987), Weissler (1985), and Wu (1995), and the references therein. The studies of Ball *et al.* (1991), Deng *et al.* (1992), Olmstead & Roberts (1996), and Straughan *et al.* (1987) address partial differential equations which contain non-local non-linearities. The articles of Ball *et al.* (1991) and Deng *et al.* (1992) develop a rigorous analysis which describes the evolutionary behavior of a solution in the neighbourhood of the blow up time, or as time becomes very large. The paper of Straughan *et al.* (1987) is a computational one based on finite elements, while the work of Olmstead & Roberts (1996) is reviewed in section 5.2 of this monograph. Also of interest is the numerical work on blowing up solutions to the Constantin *et al.* (1985) model of the vorticity equation by Stewart & Geveci (1992), and De

Gregorio's (1990) work on extending this model. Novel work on sectorial and exterior domains has been presented by Bandle & Levine (1989a, 1989b), and Levine & Meyer (1990), the latter work, like that of Levine *et al.* (1990), concentrating on critical exponents of the non-linearities. Degenerate parabolic equations where the diffusion coefficient may tend to zero have been studied by Junning (1993), Levine & Sacks (1984) and by Morro & Straughan (1983), and interesting arguments employing ideas of potential wells were utilized by Levine & Smith (1986, 1987). Keller (1957) and Levine (1974) have considered finite time blow-up in equations which possess singular coefficients. Also, the optimal control of the blow-up time in a diffusion process is investigated in Barron *et al.* (1996), where other references to this topic may be found. The paper of Etheridge (1996) is an interesting one which applies probability theory arguments to investigate the nature of the (spatial) blow-up set for a forced parabolic equation.

We should point out that we are not primarily covering the area of blow-up of derivatives in hyperbolic systems, i.e. the evolutionary behaviour of acceleration waves, and the associated theory of shock wave development. However, this is also an important area where derivatives blow-up in a finite time. Excellent accounts of this subject may be found in Chen (1973), Dafermos (1985), and Whitham (1974). Dafermos' (1985) article is a lucid account of several areas in mechanics and is very intuitive and illustrative. Dafermos (1986) deals rigorously with singularity formation in a viscoelastic material which has a fading memory while Dafermos & Hsaio (1986) consider similar questions in non-linear thermoelasticity. The paper of Morro (1978) deals with the interaction of a shock wave and an acceleration wave in an elastic body and includes several references to analogous analyses in the area of continuum mechanics. The paper of Fu & Scott (1991) is a revealing one which is concerned with the development of an acceleration wave into a shock wave. By developing an exact solution using simple wave theory they are able to cast some insight into this difficult question. There are also some recent papers dealing with thermal singularities in compressible atmospheres. Short (1995), (1996) studies thermal explosion in a gas by asymptotic and numerical methods, while Goldshtein *et al.* (1996) give detailed analysis of thermal explosions in a dusty gas. These papers contain many relevant references to explosive behaviour in compressible fluids.

To define the phenomenon of blow-up in a finite time, let u be a solution to a first order in time partial differential equation, say

$$\frac{\partial u}{\partial t} = Lu, \qquad (1.1.1)$$

for some partial derivative operator L which involves spatial derivatives. Suppose this equation is defined on a spatial domain $\Omega \subseteq \mathbf{R}^N$, for some positive range of times $t > 0$. The solution to (1.1.1) will also be required to satisfy suitable boundary and initial data. The definition of blow-up in finite time is facilitated if we define a number T^* by

$T^* = \sup\{T > 0 | u(\mathbf{x}, t) \text{ is bounded in } \Omega \times [0, T), \text{ where } u \text{ satisfies (1.1.1)}\}.$

If $T^* = +\infty$ then blow-up in finite time does not occur and solutions are said to be global. If $T^* < \infty$ then

$$\limsup_{t \to T^*} \|u(t)\|_\infty = \infty$$

and one says the solution blows up at time T^*.

The notation employed throughout this book is standard in the mathematical and mechanics literature. We shall usually use Ω to denote a spatial domain, $\Omega \subseteq \mathbf{R}^N$, for some integer $N \geq 1$. In mechanics applications $N = 1, 2$ or 3. If Ω is bounded its boundary will be denoted by Γ. The notation $\| \cdot \|$ and $(\cdot, \cdot)$ refers to the norm and inner product on $L^2(\Omega)$, respectively, i.e.

$$\|f\|^2 = \int_\Omega f^2 dx = (f, f),$$

$$(f, g) = \int_\Omega fg\, dx.$$

The notation $\| \cdot \|_p$ is employed to denote the norm on $L^p(\Omega)$, $p > 1$, so

$$\|f\|_p = \left(\int_\Omega |f|^p\, dx \right)^{1/p}.$$

1.2 Methods of Establishing Non-existence and Growth of Solutions

There are many techniques which have been employed in the mathematical literature to demonstrate when a solution to a partial differential equation blows up in a finite time, or when it ceases to exist for some other reason. It is not the object of this work to present a complete review of them all. Instead, we briefly describe with the aid of simple examples some of the techniques which have proved successful. There are many other methods, such as the comparison technique, which we do not cover, although this method is discussed in Sect. 6.2 when the applied problem of chemotaxis is encountered.

1.2.1 The Concavity Method

The method of concavity is due to Levine (1973) and is a technique which has found wide application.

The basic idea is to construct a positive definite functional, $F(t)$, of the solution to a partial differential equation, and then show $F^{-\alpha}$ is a concave function of t, for some number $\alpha > 0$. In that event, F satisfies the differential inequality,

$$\frac{d^2 F^{-\alpha}}{dt^2} \leq 0, \tag{1.2.1}$$

and so by integrations of this inequality one derives the following lower bound for $F(t)$,

$$F^\alpha(t) \geq \frac{F^{\alpha+1}(0)}{F(0) - t\alpha F'(0)}. \tag{1.2.2}$$

The function $F^\alpha(t)$ is thus bounded below by a function which blows up in finite time provided $F'(0) > 0$. The blow-up must then occur before or at

$$T^* = \frac{F(0)}{\alpha F'(0)} > 0. \tag{1.2.3}$$

This argument by itself does not establish that $F(t)$ actually blows up. However, it certainly shows the solution cannot exist for all time in a classical sense.

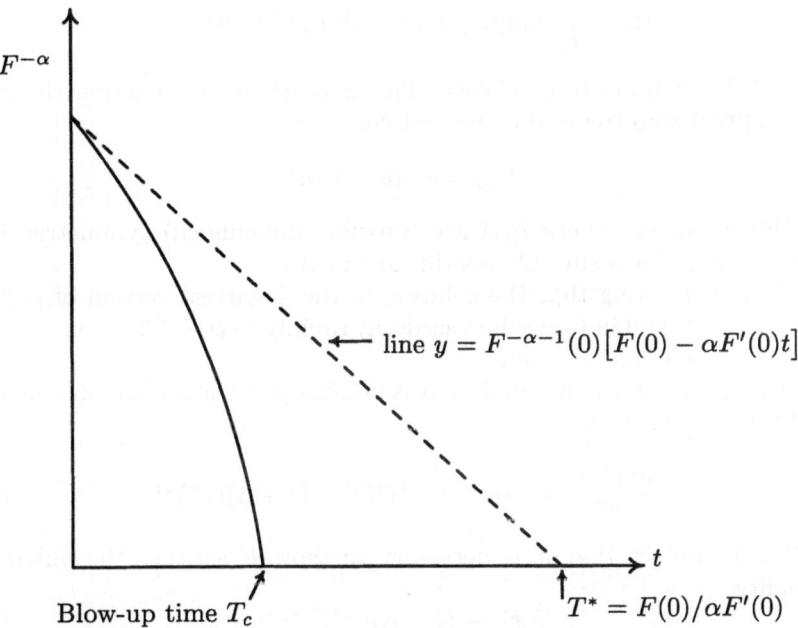

Fig. 1.1 The function $F^{-\alpha}(t)$.

The geometrical interpretation of the concavity method as may be gleaned from Fig. 1.1, is that $F^{-\alpha}(t)$ lies beneath the straight line graph, $y = F^{-\alpha-1}(0)[F(0) - \alpha t F'(0)]$. When $F'(0) > 0$ the slope of this line is negative and since $F^{-\alpha}$ is concave then it lies below this line and so cuts the t axis at a point $T_c < \infty$, provided the solution can be continued to this point.

We illustrate this method by application to the partial differential equation

$$\frac{\partial u}{\partial t} = \Delta u + u^p, \tag{1.2.4}$$

defined on $\Omega \times (0,T)$, for some $T > 0$ and Ω a bounded domain in $\mathbf{R}^N$, $N \geq 1$. The constant p has to be greater than 1 for blow-up in finite time to occur. (For an arbitrary $p > 1$ we ought to replace u^p by $u|u|^{p-1}$ in (1.2.4).) The solution u is subject to the boundary condition

$$u = 0, \qquad \text{on } \Gamma, \tag{1.2.5}$$

where Γ is the boundary of Ω, and is also required to satisfy the initial data

$$u(x,0) = u_0(x), \qquad x \in \Omega. \tag{1.2.6}$$

A suitable function $F(t)$ to employ for (1.2.4)–(1.2.6) is

$$F(t) = \int_0^t \|u\|^2 d\eta + (T - t)\|u_0\|^2 + \beta(t + \tau)^2, \tag{1.2.7}$$

where $T, \beta, \tau > 0$ are to be chosen. This is basically the function chosen by Levine (1973) who treats the abstract equation

$$Pu_t = -Au + \mathcal{F}(u),$$

in a Hilbert space, where P, A are (possibly unbounded) symmetric linear operators and $\mathcal{F}$ is a suitable non-linear function.

It is worth noting that the solution to the linearised version of (1.2.4) - (1.2.6) is stable and indeed solutions decay rapidly to zero. Thus, any blow-up is due to the non-linear term.

To apply the concavity method it is necessary to show F satisfies inequality (1.2.1). Thus, since

$$\frac{d^2 F^{-\alpha}}{dt^2} = -\alpha F^{-\alpha-2}\left[FF'' - (1+\alpha)(F')^2\right], \tag{1.2.8}$$

for $F > 0$, we see that it is necessary to show F satisfies the differential inequality

$$FF'' - (1+\alpha)(F')^2 \geq 0. \tag{1.2.9}$$

Thus, we differentiate (1.2.7) to obtain

$$\frac{dF}{dt} = 2\int_0^t (u, u_\eta)d\eta + 2\beta(t + \tau), \tag{1.2.10}$$

where $(\cdot, \cdot)$ denotes the inner product on $L^2(\Omega)$. To proceed we observe that by substitution for u_η from the partial differential equation (1.2.4) and by integration, F' may be rewritten as

$$\frac{dF}{dt} = -2\int_0^t \|\nabla u\|^2 d\eta + 2\int_0^t \int_\Omega u^{p+1} dx\, d\eta + 2\beta(t+\tau). \qquad (1.2.11)$$

This expression is differentiated and with a little rearrangement and integration by parts we find

$$F'' = 4\int_0^t (\Delta u, u_\eta) d\eta - 2\|\nabla u_0\|^2 + 2\int_\Omega u^{p+1} dx + 2\beta. \qquad (1.2.12)$$

The term Δu is substituted using (1.2.4) to then find

$$F'' = 4\int_0^t \|u_\eta\|^2 d\eta - \frac{4}{p+1}\int_\Omega u^{p+1} dx + \frac{4}{p+1}\int_\Omega u_0^{p+1} dx$$

$$- 2\|\nabla u_0\|^2 + 2\int_\Omega u^{p+1} dx + 2\beta.$$

In anticipation of forming the left hand side of inequality (1.2.9) this equation is rewritten in the following manner,

$$F'' = 4(\alpha+1)\left[\int_0^t \|u_\eta\|^2 d\eta + \beta\right] - 4\alpha \int_0^t \|u_\eta\|^2 d\eta - 2(2\alpha+1)\beta$$

$$+ 2\left(\frac{p-1}{p+1}\right)\int_\Omega u^{p+1} dx$$

$$+ 4\left[\frac{1}{p+1}\int_\Omega u_0^{p+1} dx - \frac{1}{2}\|\nabla u_0\|^2\right]. \qquad (1.2.13)$$

We now form the left hand side of inequality (1.2.9) using (1.2.13), (1.2.7), and (1.2.10) to find

$$FF'' - (1+\alpha)(F')^2 = 4(\alpha+1)S^2 + 4(\alpha+1)(T-t)\|u_0\|^2\left(\int_0^t \|u_\eta\|^2 d\eta + \beta\right)$$

$$+ F\left\{-4\alpha \int_0^t \|u_\eta\|^2 d\eta - 2(2\alpha+1)\beta + 2\left(\frac{p-1}{p+1}\right)\int_\Omega u^{p+1} dx\right.$$

$$\left. + 4\left[\frac{1}{p+1}\int_\Omega u_0^{p+1} dx - \frac{1}{2}\|\nabla u_0\|^2\right]\right\}, \qquad (1.2.14)$$

where the quantity S^2, which is non-negative by virtue of the Cauchy–Schwarz inequality, is defined by

$$S^2 = \left[\int_0^t \|u\|^2 d\eta + \beta(t+\tau)^2\right]\left[\int_0^t \|u_\eta\|^2 d\eta + \beta\right]$$

$$- \left[\int_0^t (u, u_\eta) d\eta + \beta(t+\tau)\right]^2.$$

Note that by using the partial differential equation (1.2.4) and integrating by parts we can show that

$$\int_0^t \|u_\eta\|^2 d\eta = \int_0^t (u_\eta, \Delta u + u^p) d\eta,$$

$$= -\frac{1}{2}\|\nabla u\|^2 + \frac{1}{2}\|\nabla u_0\|^2$$

$$+ \frac{1}{p+1}\int_\Omega u^{p+1} dx - \frac{1}{p+1}\int_\Omega u_0^{p+1} dx. \quad (1.2.15)$$

The first two terms on the right hand side of (1.2.14) are non-negative and we discard them. Then (1.2.15) is used to replace the first term in the curly brackets on the right hand side of (1.2.14), and we thus obtain

$$FF'' - (1+\alpha)(F')^2 \geq 2\alpha F\|\nabla u\|^2 + \left[2\left(\frac{p-1}{p+1}\right) - \frac{4\alpha}{(p+1)}\right]F\int_\Omega u^{p+1} dx$$

$$+ F\left[\frac{4(1+\alpha)}{(p+1)}\int_\Omega u_0^{p+1} dx - 2(1+\alpha)\|\nabla u_0\|^2 - 2(2\alpha+1)\beta\right].$$

We now select α to make the second term on the right disappear, i.e. we choose

$$\alpha = \frac{p-1}{2} > 0. \quad (1.2.16)$$

Upon making this selection one is left with the inequality

$$FF'' - (1+\alpha)(F')^2 \geq 2\alpha F\|\nabla u\|^2 - 2\beta p F$$

$$+ 2(p+1)F\left[\frac{1}{p+1}\int_\Omega u_0^{p+1} dx - \frac{1}{2}\|\nabla u_0\|^2\right]. \quad (1.2.17)$$

Suppose now the initial data is such that the following inequality is satisfied,

$$\int_\Omega u_0^{p+1} dx > \left(\frac{p+1}{2}\right)\|\nabla u_0\|^2. \quad (1.2.18)$$

We then pick β so small that it satisfies the restriction

$$\frac{1}{p}\int_\Omega u_0^{p+1} dx - \left(\frac{p+1}{2p}\right)\|\nabla u_0\|^2 > \beta. \quad (1.2.19)$$

Upon selecting β in this manner it immediately follows from (1.2.17) that inequality (1.2.9) holds and, therefore, F satisfies (1.2.1). Thus the solution u to (1.2.4)–(1.2.6) cannot exist beyond T^* given by (1.2.3), in a classical sense. In the present case (1.2.3) assumes the form

$$T^* = \frac{T\|u_0\|^2 + \beta\tau^2}{\beta\tau(p-1)}.$$

At the outset the number T in the definition of F in (1.2.7) should be chosen so that

$$T \geq T^* = \frac{T\|u_0\|^2 + \beta\tau^2}{\beta\tau(p-1)},$$

and thus we must choose T to satisfy

$$T \geq \frac{\beta\tau^2}{\beta\tau(p-1) - \|u_0\|^2}. \tag{1.2.20}$$

The coefficient β has already been restricted by (1.2.19), but we may now choose τ so large that

$$\tau > \frac{\|u_0\|^2}{\beta(p-1)}, \tag{1.2.21}$$

and simultaneously such that (1.2.20) is satisfied.

Thus, the solution to (1.2.4)–(1.2.6) must blow up in F measure at or before T^*, or cease to exist due to lack of regularity. In this case the solution does, in fact, blow up.

There are several interesting extensions of the concavity method. For example, Levine & Payne (1974a,1974b) have shown how it may lead to blow-up in finite time when the partial differential equation is linear but the boundary conditions are non-linear. Levine (1974) has shown how to apply the concavity technique when the coefficients in the partial differential equation are singular, by deriving a non-existence theorem for a solution to a nonlinear Euler–Poisson–Darboux equation. Also, Kalantarov & Ladyzhenskaya (1977) have shown how to extend the non-existence argument to the generalised concavity inequality

$$FF'' - (1+\alpha)(F')^2 \geq -aF^2 - bFF'. \tag{1.2.22}$$

When $b = 0$ it is not difficult to see why this works. For then, by multiplying (1.2.22) by $-\alpha F^{-\alpha-2}$, one demonstrates that

$$(F^{-\alpha})'' \leq a\alpha F^{-\alpha}.$$

Put $\omega = \sqrt{a\alpha} > 0$. Then with $X = F^{-\alpha}$, this inequality may be arranged in such a way as to see

$$\frac{d}{dt}\left[e^{-\omega t}\left(\frac{dX}{dt} + \omega X \right) \right] \leq 0. \tag{1.2.23}$$

Put $X_0' + \omega X_0 = Q_0$ and require that $Q_0 < 0$. Then from inequality (1.2.23) we may establish by integration that

$$\frac{1}{X(t)} = F^\alpha(t) \geq \frac{1}{X_0 e^{-\omega t} + (Q_0/2a)(e^{\omega t} - e^{-\omega t})}.$$

The right hand side of this expression blows up when

$$T^* = \frac{1}{2\omega} \log\left(1 - \frac{2\omega X_0}{Q_0} \right),$$

$$= \frac{1}{2\omega} \log\left[1 + \frac{2\omega F(0)}{\alpha F'(0) - \omega F(0)} \right].$$

Thus, blow-up may be established provided

$$F'(0) > \frac{\omega}{\alpha} F(0) = \sqrt{\frac{a}{\alpha}}\, F(0)\,.$$

Kalantarov & Ladyzhenskaya (1977) give many examples of partial differential equations for which their arguments lead to blow-up in finite time.

Another illustration of the method of concavity is given in the book by Flavin & Rionero (1995), p. 108–115. To illustrate the method, these writers describe how the concavity technique may be applied to establish non-existence for a solution to a non-linear Schrödinger equation, and to establish non-existence for a solution to the equations for a non-linear elastic rod. They also show, Flavin & Rionero (1995), p. 305, how to establish global non-existence for a solution to an equation of Monge–Ampère type.

1.2.2 The Eigenfunction Method

To illustrate this method we again use the example of the boundary initial value problem (1.2.4)–(1.2.6). Suppose now that $u(x,t) > 0$ for $x \in \Omega$. Let ϕ be the first eigenfunction of the membrane problem for Ω, i.e. ϕ satisfies the boundary value problem,

$$\begin{aligned} \Delta\phi + \lambda_1\phi &= 0, &\quad \text{in } \Omega, \\ \phi &= 0, &\quad \text{on } \Gamma, \end{aligned} \tag{1.2.24}$$

where λ_1 is the lowest eigenvalue. It is known from the Courant nodal line theorem that $\phi > 0$ in Ω and this fact is important here.

To describe the eigenfunction method we construct the function F as

$$F(t) = \int_\Omega \phi(x)u(x,t)\,dx = (\phi, u). \tag{1.2.25}$$

Differentiate, use the partial differential equation (1.2.4) and the boundary condition (1.2.5) to see that

$$\begin{aligned} F' &= (\phi, u_t), \\ &= (\phi, \Delta u + u^p), \\ &= -(\nabla\phi, \nabla u) + (\phi, u^p), \\ &= (\Delta\phi, u) + (\phi, u^p), \\ &= -\lambda_1(\phi, u) + (\phi, u^p), \end{aligned} \tag{1.2.26}$$

where in arriving at the last line (1.2.24) has been employed. Next, observe that from Hölder's inequality

$$\int_\Omega \phi u\,dx \le \left(\int_\Omega \phi\,dx\right)^{(p-1)/p} \left(\int_\Omega \phi u^p\,dx\right)^{1/p}.$$

Suppose ϕ is normalised so that $\int_\Omega \phi \, dx = 1$. Then,

$$\int_\Omega \phi u^p \, dx \geq \left(\int_\Omega \phi u \, dx \right)^p .$$

Then, use of this inequality in (1.2.26) leads to

$$\frac{dF}{dt} \geq -\lambda_1 F + F^p. \tag{1.2.27}$$

The proof now proceeds by contradiction. Assume a solution to (1.2.4)–(1.2.6) exists for all $t > 0$. We suppose the initial data is such that

$$F^p(0) - \lambda_1 F(0) > 0. \tag{1.2.28}$$

Then by continuity there exists a $t_1 > 0$ such that the function Q defined by

$$Q(F) = F^p - \lambda_1 F,$$

is positive on $[0, t_1)$. Since $Q' > 0$ on $[0, t_1)$,

$$Q\big(F(t_1)\big) \geq Q\big(F(0)\big) > 0.$$

Hence $t_1 = \infty$ or the limit of the existence interval of the solution. Thus, the right hand side of (1.2.27) is positive and so we may separate variables to find

$$t \leq \int_{F(0)}^{F(t)} \frac{dF}{F^p - \lambda_1 F} \leq \int_{F(0)}^{\infty} \frac{dF}{F^p - \lambda_1 F} < \infty . \tag{1.2.29}$$

Clearly t is finite and thus a contradiction is achieved. Hence, the solution cannot exist for all positive time.

There are many extensions of the above idea. Levine (1975) has shown that the restriction of positive solutions is not necessary and has extended the argument in various other directions. Ni et $al.$ (1984) also use the eigenfunction method in an interesting and rigorous manner. The eigenfunction method is ideally suited to treating any convex non-linearity, e.g. it is eminently suitable to apply to such a partial differential equation as

$$\frac{\partial u}{\partial t} = \Delta u + f(u),$$

where $f(u)$ is a convex function of u. An example of a such a convex function $f(u)$ which occurs in explosion dynamics is $f(u) = e^u$. In fact, Jones (1973, 1974) studies the solution to the thermal convection system

$$\frac{\partial u_i}{\partial t} + \frac{1}{Pr} u_j \frac{\partial u_i}{\partial x_j} = -\frac{\partial p}{\partial x_i} + \Delta u_i + R\theta \delta_{i3},$$

$$\frac{\partial u_i}{\partial x_i} = 0, \tag{1.2.30}$$

$$Pr \frac{\partial \theta}{\partial t} + u_i \frac{\partial \theta}{\partial x_i} = \Delta \theta + \delta e^{\theta/(1+\beta\theta)}.$$

In the above system of partial differential equations u_i, θ, p are the velocity, temperature, and pressure, respectively. The constants Pr, R, δ and β are the Prandtl number, the Rayleigh number, and two constants measuring the size of the chemical reaction. This system is used to model convection in a heat transfer process when exothermic chemical reactions occur in the fluid body. In essence, the system of partial differential equations (1.2.30) is essentially the Navier–Stokes equations with the Boussinesq approximation used in the body force to give the $R\theta\delta_{i3}$ term, and with the temperature equation forced by an exponential reaction term. Jones (1973) studies the stability of a solution to the system (1.2.30) while Jones (1974) investigates how the velocity and temperature evolve in time, by a numerical method. Jones (1974), actually, works with $\beta = 0$. This is thus a fluid dynamic analogue of the single partial differential equation case with an exponential force. Logan (1994) briefly considers applications of the eigenfunction method in his book. He refers [Logan (1994), p. 296] to the partial differential equation

$$\frac{\partial u}{\partial t} = \frac{\partial^2 u}{\partial x^2} + \delta e^u,$$

together with zero boundary and initial data as the Gelfand problem.

1.2.3 Explicit Inequality Methods

Into this category fall a whole plethora of techniques. The basic idea is to select a positive-definite functional and then show it is bounded below by a function of t which becomes singular in finite time. We illustrate a typical procedure by employing a partial differential equation which in the mathematical literature is referred to as the porous medium equation, namely,

$$\frac{\partial u}{\partial t} = \Delta(u^m), \qquad (1.2.31)$$

for $m > 0$ a constant. We shall study this equation backward in time as in Levine & Payne (1974a), and so we reverse time and consider instead of (1.2.31) the partial differential equation

$$\frac{\partial u}{\partial t} = -\Delta(u^m), \qquad (1.2.32)$$

for some $m > 1$, defined on the domain $\Omega \times (0, T)$, for some $T < \infty$ with Ω a bounded region in $\mathbf{R}^N$. On the boundary, Γ, of Ω we suppose

$$u = 0, \qquad \text{on } \Gamma. \qquad (1.2.33)$$

Levine & Payne (1974a) investigate the one-dimensional version of (1.2.32) but with the spatial domain being all of the real line. This is a harder problem. In fact, the porous medium equation has been employed in mathematical biology, in particular to describe the dispersal of insects and other animals, see Murray (1990), pp. 238–241. Equation (1.2.31) may be written as

$$\frac{\partial u}{\partial t} = m\nabla(u^{m-1}\nabla u),$$

from which we recognize it as effectively a diffusion equation with a non-linear diffusion coefficient $\kappa = mu^{m-1}$. The solution dependence of the diffusivity removes the sometimes undesirable effect associated with the solution to the standard linear diffusion equation, i.e. κ constant, where a disturbance travels with infinite speed. The solution dependent diffusivity allows the solution to (1.2.31) to spread out like a wave in space–time and is, therefore, suitable to model the distribution of an insect population such as a swarm of mosquitoes. For further details of diffusion in mathematical biology and its application to the spatio-temporal spread of insect populations the reader is referred to the excellent book by Murray (1990), and the references therein. Logan (1994), pp. 23–25, shows how the porous medium equation may be derived from the fluid dynamical equations of liquid flow in a porous solid matrix. An interesting mathematical analysis of localization in both space and time for a system of equations which describes flow in a porous medium may be found in Antontsev & Diaz (1991). This paper contains several references to related work on systems based on Darcy's law, the equation of continuity, and the equation expressing pressure balance in the various phases present in the layer.

To establish global non-existence of a solution to (1.2.32), suppose $m > 1$ and then multiply this equation by u to find

$$\frac{d}{dt}\frac{1}{2}\|u\|^2 = m\int_{\Omega} u^{m-1}|\nabla u|^2 dx,$$

$$= \frac{4m}{(m+1)^2}\|\nabla u^{(m+1)/2}\|^2.$$

The Poincaré inequality for a function $f(x)$ which is defined on Ω and zero on Γ, i.e.

$$\|\nabla f\|^2 \geq \lambda_1 \|f\|^2,$$

where λ_1 is the lowest eigenvalue of the membrane problem (1.2.24), is now used on the right hand side to find

$$\frac{d}{dt}\|u\|^2 \geq \frac{8m\lambda_1}{(m+1)^2}\int_{\Omega} u^{m+1} dx. \tag{1.2.34}$$

Finally, Hölder's inequality in the form

$$\int_{\Omega} u^2\, dx \leq [M(\Omega)]^{(m-1)/(m+1)}\left(\int_{\Omega} u^{m+1} dx\right)^{2/(m+1)},$$

where $M(\Omega)$ is the Lebesgue measure of Ω, employed in (1.2.34) yields,

$$\frac{d}{dt}\|u\|^2 \geq \frac{8m\lambda_1}{(m+1)^2 M^{(m-1)/2}} \|u\|^{m+1}. \qquad (1.2.35)$$

This inequality may be integrated to find

$$\|u(t)\| \geq \frac{1}{\left(\|u_0\|^{m-1} - At\right)^{1/(m-1)}}, \qquad (1.2.36)$$

where the constant A is given by

$$A = \frac{4m(m-1)\lambda_1}{(m+1)^2 M^{(m-1)/2}}.$$

The right hand side of (1.2.36) blows up at $T^* = \|u_0\|^{m-1}/A$, and so this represents an upper bound for the existence interval for u.

It is of interest to observe that (1.2.4)–(1.2.6) may also be treated by a similar method to that of this section. To see this note that with $F(t) = \|u(t)\|^2$,

$$F' = -2\|\nabla u\|^2 + 2\int_\Omega u^{1+p}\, dx.$$

Also, by multiplication of (1.2.4) by u_t and integration,

$$2\int_0^t \|u_\eta\|^2 d\eta + \|\nabla u\|^2 = \frac{2}{p+1}\int_\Omega u^{p+1}\, dx + \|\nabla u_0\|^2 - \frac{2}{p+1}\int_\Omega u_0^{p+1}\, dx.$$

Use the above equation to substitute for $\|\nabla u\|^2$ in the equation for F' and then note that from Hölder's inequality

$$\int_\Omega u^{p+1}\, dx \geq F^{(p+1)/2}\left[M(\Omega)\right]^{-(p-1)/2},$$

where $M(\Omega)$ is the measure of Ω. Use of this inequality leads to

$$\begin{aligned}
F' &= \frac{2(p-1)}{p+1}\int_\Omega u^{p+1}\, dx + 4\int_0^t \|u_\eta\|^2 d\eta \\
&\quad + 2\left[\frac{2}{p+1}\int_\Omega u_0^{p+1}\, dx - \|\nabla u_0\|^2\right] \\
&\geq \frac{2(p-1)}{(p+1)\left[M(\Omega)\right]^{(p-1)/2}} F^{(p+1)/2} + 2\left[\frac{2}{p+1}\int_\Omega u_0^{p+1}\, dx - \|\nabla u_0\|^2\right].
\end{aligned}$$

Thus, provided (1.2.18) holds

$$F' \geq \frac{2(p-1)}{(p+1)\left[M(\Omega)\right]^{(p-1)/2}} F^{(p+1)/2}.$$

An estimate analogous to (1.2.36) follows easily from this inequality and hence an upper bound for the existence interval may be determined.

1.2.4 The Multi-Eigenfunction Method

This method works when a convective non-linear term is present and this is important because non-linear convection terms arise frequently in mechanics. To illustrate this consider the boundary initial value problem

$$\frac{\partial u}{\partial t} + u\frac{\partial u}{\partial x} = \frac{\partial^2 u}{\partial x^2} + \beta u^{2+\delta}, \qquad x \in (0,1), \ t > 0,$$
$$u(0,t) = u(1,t) = 0, \qquad \forall t \geq 0, \tag{1.2.37}$$
$$u(x,0) = u_0(x),$$

for some $\beta > 0$ and any $\delta > 0$. This problem is illustrated in Straughan (1992), Sect. 2.6, but we include it here for clarity.

Denote by ϕ the first eigenfunction of the membrane problem for the spatial region, in this case $\{x | x \in (0,1)\}$, so that we may take

$$\phi = \sin \pi x.$$

Define a function $F(t)$ now by

$$F(t) = (\phi^n, u),$$

where $n > 1$ is to be chosen. We differentiate F and use (1.2.37) to see that

$$\frac{dF}{dt} = (\phi^n, u_t),$$
$$= (\phi^n, -uu_x + u_{xx} + \beta u^{2+\delta}),$$
$$= -\pi^2 nF + n(n-1)\int_0^1 \phi^{n-2}\phi_x^2 u\, dx$$
$$+ \beta \int_0^1 \phi^n u^{2+\delta} dx + \frac{1}{2}n \int_0^1 \phi^{n-1}\phi_x u^2\, dx. \tag{1.2.38}$$

The second term on the right is discarded and ϕ_x is bounded below by $-\pi$ to arrive at

$$\frac{dF}{dt} \geq -\pi^2 nF + \beta \int_0^1 \phi^n u^{2+\delta} dx - \frac{1}{2}n\pi \int_0^1 \phi^{n-1} u^2\, dx. \tag{1.2.39}$$

From Hölder's inequality

$$\int_0^1 \phi^{2n/(2+\delta)} u^2\, dx \leq \left(\int_0^1 1\, dx\right)^{\delta/(2+\delta)} \left(\int_0^1 \phi^n u^{2+\delta}\, dx\right)^{2/(2+\delta)},$$

and so we pick $n = 1 + 2/\delta$ to see that

$$\int_0^1 \phi^{1+2/\delta} u^{2+\delta}\, dx \geq \left(\int_0^1 \phi^{2/\delta} u^2\, dx\right)^{(2+\delta)/2}. \tag{1.2.40}$$

We next use (1.2.40) in (1.2.39) to derive

$$\frac{dF}{dt} \geq -\left(\frac{\delta+2}{\delta}\right)\pi^2 F + \beta\left(\int_0^1 \phi^{2/\delta}u^2\,dx\right)^{(2+\delta)/2}$$
$$-\frac{1}{2}\pi\left(\frac{\delta+2}{\delta}\right)\int_0^1 \phi^{2/\delta}u^2\,dx. \qquad (1.2.41)$$

To continue we use the Cauchy–Schwarz inequality to show

$$\int_0^1 \phi^{2/\delta}u^2\,dx \geq \frac{\left(\int_0^1 \phi^{1+2/\delta}u\,dx\right)^2}{\int_0^1 \phi^{2+2/\delta}\,dx},$$

$$\geq \left(\int_0^1 \phi^{1+2/\delta}u\,dx\right)^2. \qquad (1.2.42)$$

If we define $G(t)$ by

$$G(t) = \int_0^1 \phi^{2/\delta}u^2\,dx,$$

then (1.2.42) is

$$G(t) \geq F^2(t).$$

Thus, from (1.2.41) we derive

$$\frac{dF}{dt} \geq -\left(\frac{\delta+2}{\delta}\right)\pi^2 G^{1/2} + \beta G^{(2+\delta)/2} - \frac{1}{2}\pi\left(\frac{\delta+2}{\delta}\right)G. \qquad (1.2.43)$$

Define the function $R(F)$ by

$$R(F) = \beta F^{2+\delta}(t) - \frac{1}{2}\pi\left(\frac{\delta+2}{\delta}\right)F^2(t) - \left(\frac{\delta+2}{\delta}\right)\pi^2 F(t). \qquad (1.2.44)$$

At this point we must assume the initial data function $u_0(x)$ is such that

$$R\big(F(0)\big) > 0.$$

By continuity there exists a $t_1 > 0$ such that for $t \in [0, t_1)$,

$$R\big(F(t)\big) > 0.$$

By computation of $R'(F)$ we find that for $t \in [0, t_1)$, $R'\big(F(t)\big) > 0$, i.e. R is increasing on this interval. Thus, since $G^{1/2} \geq F$,

$$R\big(G^{1/2}(t)\big) \geq R\big(F(t)\big), \qquad t \in [0, t_1).$$

Hence, from (1.2.43) we find for $t \in [0, t_1)$,

$$\frac{dF}{dt} \geq R(F). \qquad (1.2.45)$$

Both F' and $R(F)$ are increasing on $[0, t_1)$ and, therefore, $R\big(F(t_1)\big) \neq 0$. It follows that t_1 is either infinity or the limit of existence of the solution. We may now separate variables in (1.2.45) to derive

$$t_1 \leq \int_{F(0)}^{F(t)} \frac{dF}{R(F)} \leq \int_{F(0)}^{\infty} \frac{dF}{R(F)} < \infty.$$

This leads to a contradiction if the case of infinite t_1 is considered. Hence, we deduce the solution to (1.2.37) ceases to exist globally in a finite time.

In Levine *et al.* (1989) it is shown that the solution to (1.2.37) does in fact blow up in a finite time.

In connection with the topic of this subsection it is worth pointing out that the effect of convection terms on the development of boundary layers in parabolic equations with small diffusion is studied in some detail in Howes (1986a,1986b). Howes (1986a) studies the stability of boundary layer solutions to the boundary initial value problem

$$\frac{\partial u}{\partial t} + a(x, u)\frac{\partial u}{\partial x} + b(x, u) = \epsilon \frac{\partial^2 u}{\partial x^2}, \qquad (x, t) \in (0, 1) \times (0, \infty),$$

$$u(x, 0, \epsilon) = \psi(x, \epsilon), \qquad x \in (0, 1),$$

$$u(0, t, \epsilon) = \alpha, \qquad u(1, t, \epsilon) = \beta, \qquad t \geq 0.$$

In Howes (1986b) the boundary layer structure is examined for a solution to the partial differential equation

$$\frac{\partial u}{\partial t} + \sum_{i=1}^{N} a_i(\mathbf{x}, t, u)\frac{\partial u}{\partial x_i} + b(\mathbf{x}, t, u) = \epsilon \sum_{i=1}^{N} \frac{\partial^2 u}{\partial x_i \partial x_i},$$

for $\epsilon \to 0$.

The complete solution structure, including that of the boundary layer as well as that of the solution elsewhere in the spatial domain, of a system of partial differential equations with convection terms, is displayed by numerical calculations in Sect. 2.5 of this monograph.

1.2.5 Logarithmic Convexity

While the main thrust of this monograph is to consider problems where the solution blows up in a finite time, there are many physical situations where a very rapid growth such as exponential is every bit as important. Hence, we now describe a technique which can lead to exponential growth of a solution.

The technique we shall describe is called logarithmic convexity. The method of logarithmic convexity is, in some sense, the limit of the concavity technique as $\alpha \to 0$, and so it does not yield a finite time blow-up estimate. Logarithmic convexity is, in fact, a method capable of producing many other useful results rather than just growth estimates and the use of this technique in improperly posed problems to establish continuous dependence in many situations is considered in detail in the books of Ames & Straughan (1997) and Payne (1975). Payne (1971, 1993) also considers this method in the context of the Navier–Stokes equations. We here illustrate the technique of logarithmic convexity by showing how its use can lead to a growth estimate for a solution to a linear partial differential equation. The partial differential equation in question is

$$\frac{\partial u}{\partial t} = \Delta u + ku, \tag{1.2.46}$$

on the region $\Omega \times (0, \infty)$, $\Omega \subset \mathbf{R}^N$, Ω bounded, where k is a positive constant. The boundary condition we consider is

$$u = 0, \qquad \text{on } \Gamma,$$

with Γ being the boundary of Ω. The initial data is given as $u_0(x)$.

If k is sufficiently small then it is well known the solution u to (1.2.46) subject to zero boundary conditions decays to zero exponentially fast. Thus, conditions on k guaranteeing growth are also of use.

We construct a function whose logarithm is a convex function of t. So, let F be defined by

$$F(t) = \|u(t)\|^2. \tag{1.2.47}$$

Differentiation yields

$$F' = 2(u, u_t),$$

and then

$$F'' = 2\|u_t\|^2 + 2(u, u_{tt}),$$
$$= 2\|u_t\|^2 + 2(u, \Delta u_t + ku_t),$$

upon substitution from the partial differential equation (1.2.46). Two integrations by parts and resubstitution from (1.2.46) lead to

$$F'' = 2\|u_t\|^2 + 2(\Delta u, u_t) + 2k(u, u_t),$$
$$= 2\|u_t\|^2 + 2(u_t, u_t - ku) + 2k(u, u_t),$$
$$= 4\|u_t\|^2.$$

Therefore, by the equation just derived and the Cauchy–Schwarz inequality

$$FF'' = 4\|u\|^2\|u_t\|^2$$
$$\geq 4(u, u_t)^2$$
$$= (F')^2.$$

This means that

$$FF'' - (F')^2 \geq 0.$$

Thus, for $F > 0$, we find by division of F^2,

$$\frac{d^2}{dt^2} \log F \geq 0, \tag{1.2.48}$$

i.e. $F(t)$ is a logarithmically convex function of t.

From integration of (1.2.48) we find that

$$F(t) \geq F(0) \exp \left[\frac{F'(0)t}{F(0)} \right].$$ (1.2.49)

(From this inequality we see that $F'(0) > 0$ ensures $F(t) > 0$ for all $t \geq 0$ and thus the division by F necessary to derive inequality (1.2.48) is justified.) Inequality (1.2.49) may be rewritten as,

$$\|u(t)\|^2 \geq \|u_0\|^2 \exp \left[\frac{2(u(0), u_t(0))}{\|u_0\|^2} t \right].$$ (1.2.50)

It is seen that growth in (1.2.50) requires

$$(u(0), u_t(0)) > 0,$$

or, using the partial differential equation (1.2.46), this condition may be seen to be equivalent to

$$k\|u_0\|^2 > \|\nabla u_0\|^2.$$ (1.2.51)

Thus, if k and the initial data satisfy inequality (1.2.51), the solution must grow at least exponentially in $L^2(\Omega)$ norm.

Many growth results for abstract differential equations are derived by means of the logarithmic convexity method by Levine (1972).

1.3 Finite Time Blow-Up Systems with Convection

The mathematical models we now mention may often be regarded as simple models for certain types of gas dynamic or fluid dynamical behaviour, see e.g. Cox & Mortell (1983), Cox (1991), Budd *et al.* (1994). Models for non-linear thermal convection incorporating penetrative convection effects may often involve non-linearities not dissimilar to those discussed in the partial differential equations examples which follow, cf. Straughan (1993). The phenomenon of penetrative convection arises when a stable layer of fluid is surrounded by one or more unstable layers of fluid. The convective motion in the unstable layer(s) tends to induce motion in the adjacent stable layer which in turn becomes convectively unstable. Such motions have wide mundane application, in oceanography, motions in the interiors of planets, cloud physics, and many more situations, and so are physically important. A typical such example is when the Earth is heated by the Sun in the morning. The lower layer of air becomes hotter than that above due to radiation heating of the Earth's surface and then begins to move convectively. This has the effect of penetrating into the higher air layers. A simple example of penetrative convection is when a horizontal layer of water is such that the bottom of the layer is maintained at a temperature of 0°C while the top of the layer is maintained at a temperature $T_u > 4°C$. Since water has a density maximum at approximately $4°C$, the water in the layer below $4°C$ is in an unstable configuration due to the downward effect of gravity. Thus motion may ensue in the lower part of the layer and penetrate into the upper layer. The density relation used to describe such penetrative convection effects is typically

$$\rho = \rho_0\big(1 - \alpha(T - 4)^2\big),$$

where ρ is density, T is temperature, and ρ_0, α are constants. Alternatively, higher order polynomial relations $\rho(T)$ are sometimes used; a complete account of this may be found in the monograph of Straughan (1993). The upshot of this is that polynomial functions occur naturally as forcing terms in the parabolic-like partial differential equations which govern the fluid dynamics. Hence, a study of partial differential equations with polynomial forces is useful. In fact, the penetrative convection phenomenon can lead to novel effects in fluid dynamics. For example, Straughan & Walker (1997) have discovered that in penetrative convection with temperature and *two* different salt concentrations, convective instability may arise at the same critical Rayleigh numbers, *simultaneously* by a stationary convection mechanism and also by an oscillatory convection mechanism, although the cell sizes are different.

To begin to understand the importance of a non-linear convective term on the effect of blow-up in finite time we recollect a simple example. Let u be a non-negative solution to the boundary initial value problem

$$\frac{\partial u}{\partial t} + \epsilon u \frac{\partial u}{\partial x} = \frac{\partial^2 u}{\partial x^2} + u^p, \qquad x \in (0,1),\ t > 0,$$
$$u(x,0) = u_0(x),$$
$$u(0,t) = u(1,t) = 0,$$

$$(1.3.1)$$

where $\epsilon \geq 0, p > 0$, are constants. When $\epsilon = 0$, i.e. the non-linear convective term is absent, then blow-up in finite time of u may occur when $p > 1$, see e.g. Levine (1973). However, when $\epsilon = 1$, i.e. convection is fully present, then blow-up in finite time does not occur for $p \leq 2$. Indeed, the solution may blow-up in finite time only when $p > 2$, see e.g. Levine *et al.* (1989). Thus, for $\epsilon = 1, 1 < p \leq 2$, the term uu_x is in some sense acting as a stabilizing ingredient.

1.3.1 Fujita-Type Problems

A classical result of Fujita (1966) concerns the solution to the initial value problem

$$\frac{\partial u}{\partial t} = \Delta u + u^p, \qquad \mathbf{x} \in \mathbf{R}^N,\ t > 0,$$
$$u(\mathbf{x},0) = u_0(\mathbf{x}), \qquad \mathbf{x} \in \mathbf{R}^N,$$

$$(1.3.2)$$

with $u_0 \geq 0, p > 0, N \geq 1$. As described in Sect. 1.1, we may define a number T^* by

$$T^* = \sup\{T > 0 | u(\mathbf{x}, t) \text{ is bounded in } \mathbf{R}^N \times [0, T), \text{ where } u \text{ satisfies } (1.3.2)\}.$$

If $T^* = +\infty$ then blow-up in finite time does not occur and solutions are said to be global whereas if $T^* < \infty$ then

$$\limsup_{t \to T^*} \|u(t)\|_\infty = \infty$$

and one says the solution blows up at time T^*. The subject of how the solution blows up, i.e. pointwise or otherwise has attracted much recent attention but we do not discuss this here. Nor do we consider the question of continuing the solution beyond T^*.

Fujita's (1966) classical paper shows that for $p \in (1, 1+2/N)$, solutions to (1.3.2) blow-up whereas if $p > 1 + 2/N$ then the solution may exist globally for u_0 suitably small. Hayakawa (1973) showed that when $p = 1 + 2/N$ the solution blows up.

The extension of Fujita-type results to systems of partial differential equations has occupied much recent attention.

Escobedo & Herrero (1991) study the system

$$\frac{\partial u}{\partial t} = \Delta u + v^p, \qquad \mathbf{x} \in \mathbf{R}^N, \ t > 0,$$
$$\frac{\partial v}{\partial t} = \Delta v + u^q, \qquad \mathbf{x} \in \mathbf{R}^N, \ t > 0, \qquad (1.3.3)$$
$$u(\mathbf{x}, 0) = u_0(\mathbf{x}), \quad v(\mathbf{x}, 0) = v_0(\mathbf{x}), \qquad \mathbf{x} \in \mathbf{R}^N,$$

with $p, q > 0$. They observe that since the spatial domain is all of $\mathbf{R}^N$ one may construct exact solutions which blow-up in finite time. Such an example they give is

$$u = \frac{A}{(T - t)^\alpha}, \qquad v = \frac{B}{(T - t)^\beta},$$
$$\alpha = \frac{p + 1}{pq - 1}, \qquad \beta = \frac{q + 1}{pq - 1},$$
$$A = (\alpha\beta^p)^{1/(pq-1)}, \qquad B = \frac{A^q}{\beta},$$

provided $pq > 1$.

In fact, Escobedo & Herrero (1991) show in the general case that for (1.3.3) if $0 < pq \leq 1$ then $T^* = \infty$, i.e. no blow-up occurs, so $pq > 1$ is necessary for blow-up.

In this monograph we are primarily interested in blow-up and so report only those results pertaining to this. We stress that most of the works referred to here also deal with the question of what parameter ranges lead to global existence.

Escobedo & Herrero (1991) show that for (1.3.3), if

$$\gamma = \max\{p, q\},$$

then $pq > 1$ and

$$\gamma > \frac{1}{2}N(pq - 1) - 1$$

guarantee blow-up, in that

$$\limsup_{t \to T^*} \|u(t)\|_\infty = \infty; \qquad \limsup_{t \to T^*} \|v(t)\|_\infty = \infty.$$

They also establish that if

$$pq > 1, \quad \gamma < \frac{1}{2}N(pq - 1) - 1$$

but

$$f(\mathbf{x}) \geq C \exp(-\alpha|\mathbf{x}|^2),$$

where $f = u_0$ or v_0, for some $\alpha, C > 0$, then blow-up occurs.

Escobedo & Herrero (1993) extend their earlier results to the case where the spatial domain $\mathbf{R}^N$ is replaced by a bounded domain $\Omega \subset \mathbf{R}^N$, with Ω having a smooth boundary Γ. On Γ, $u = v = 0$. For this situation Escobedo & Herrero (1993) show that when $pq \leq 1$ every solution is global. When $pq > 1$ they show that for a constant C large enough u, v blow-up in finite time when

$$\int_\Omega (u_0 + v_0)\phi_1 \, dx \geq C,$$

where ϕ_1 is the first eigenfunction in the membrane problem for Ω, i.e.

$$\Delta\phi_1 + \lambda\phi_1 = 0, \qquad \text{in } \Omega,$$
$$\phi_1 = 0, \qquad \text{on } \Gamma.$$

(The function ϕ_1 is assumed to be normalized, i.e.

$$\int_\Omega \phi_1 \, dx = 1,$$

and by the Courant nodal line theorem $\phi_1 > 0$.)

Escobedo & Levine (1995) investigate the problem

$$\frac{\partial u}{\partial t} = \Delta u + u^{p_1}v^{q_1}, \qquad \mathbf{x} \in D, \, t > 0,$$

$$\frac{\partial v}{\partial t} = \Delta v + u^{p_2}v^{q_2}, \qquad \mathbf{x} \in D, \, t > 0, \qquad (1.3.4)$$

$$u = v = 0, \qquad \text{on } \partial D,$$

$$u(\mathbf{x}, 0) = u_0(\mathbf{x}) \geq 0, \qquad v(\mathbf{x}, 0) = v_0(\mathbf{x}) \geq 0,$$

for constants $p_1, p_2, q_1, q_2 \geq 0$. Here D is either $\mathbf{R}^N$ or a cone in $\mathbf{R}^N$ with vertex at the origin.

They define the numbers α, β, δ by

$$\delta = (p_1 + q_1 - 1)(q_2 - q_1 - 1) + q_1(p_1 + q_1 - p_2 - q_2),$$

$$\alpha = \frac{q_2 - q_1 - 1}{\delta}, \qquad \beta = \frac{p_1 - p_2 - 1}{\delta}, \qquad \delta \neq 0.$$

Then Escobedo & Levine (1995) establish three technical results stated as theorems 5–7 in their paper. Let $D = \mathbf{R}^N$. For the blow-up case they may be interpreted as stating:

Suppose $p_2 q_1 > 0$. When $p_1 > 1$, then

$$\frac{1}{p_1 + q_1 - 1} \geq \frac{1}{2}N$$

guarantees blow-up, whereas if

$$\frac{1}{p_1 + q_1 - 1} < \frac{1}{2}N$$

blow-up may or may not occur. Suppose $\delta \neq 0$ and let

$$\nu = \max\{\alpha, \beta\}.$$

When $0 \leq p_1 \leq 1$, if $\nu \geq \frac{1}{2}N$ all non-trivial solutions blow-up; for $0 \leq \nu < \frac{1}{2}N$ blow-up may or may not occur.

If $\delta \neq 0$ and $p_2 q_1 = 0$, say $p_2 + q_1 > 0$, let

$$\rho = \min\{\alpha, \beta\}.$$

Then $\rho \geq \frac{1}{2}N$ assures blow-up. For $0 \leq \nu \leq \frac{1}{2}N$ blow-up may or may not occur. (Escobedo & Levine (1995) discuss the above results in greater detail than this.)

Escobedo & Levine (1995) also consider the case $\delta = 0$, $p_1 \in [0, 1]$. Additionally they extend all their results to the case when D is a cone in $\mathbf{R}^N$.

Fila *et al.* (1994) deals with the system

$$\frac{\partial u}{\partial t} = \delta \Delta u + v^p,$$

$$\frac{\partial v}{\partial t} = \Delta v + u^q, \tag{1.3.5}$$

where the spatial domain is $\mathbf{R}^N$ and δ, p, q are given constants with $pq > 0$, $0 \leq \delta \leq 1$. They establish precise conditions for blow-up and for global existence of solutions to system (1.3.5).

Lu & Sleeman (1993,1994) also consider questions of blow-up for systems of the form

$$\frac{\partial u}{\partial t} = k_1 \frac{\partial^2 u}{\partial x^2} + f_1(u, v),$$
$$\frac{\partial v}{\partial t} = k_2 \frac{\partial^2 v}{\partial x^2} + f_2(u, v),$$

(1.3.6)

for constants k_1, k_2, and various functions f_1, f_2. They pay particular attention to Fujita-type systems of the form (1.3.3)–(1.3.5). In Lu & Sleeman (1994) they remark that applied to (1.3.4) their results are not as sharp as those of Escobedo & Levine (1995), but they are dealing with a wider class of equations. In Lu & Sleeman (1993) they are particularly interested in the Fujita system (1.3.3) and two other systems. One of these two systems of equations they term the *Friedman–Giga* system (Friedman (1988), Friedman & Giga (1987))

$$\frac{\partial u}{\partial t} = A_1 \frac{\partial^2 u}{\partial x^2} + f_1(v),$$
$$\frac{\partial v}{\partial t} = A_2 \frac{\partial^2 v}{\partial x^2} + f_2(u),$$

(1.3.7)

defined for $x \in (-a, a)$, $t > 0$, with $A_\alpha > 0$, and zero boundary data for u, v at $x = \pm a$. Of course, initial data are given. The functions f_α can have various forms such as

$$f_1 = Ae^{\lambda v}, \quad f_2 = Be^{\mu u}, \quad A, B, \lambda, \mu > 0,$$

although several other functions f_α are possible. The other system Lu & Sleeman (1993) study is a very interesting one called the *Henon–Heiles* system; this is

$$\frac{\partial u}{\partial t} = \frac{\partial^2 u}{\partial x^2} + u + u^2 - v^2,$$
$$\frac{\partial v}{\partial t} = \frac{\partial^2 v}{\partial x^2} + v - 2uv.$$

(1.3.8)

This is investigated by Lu & Sleeman (1993) for $x \in (-a, a), t > 0$, with u, v subject to zero boundary data, and non-zero initial data prescribed. Blow-up results for solutions to both (1.3.7) and (1.3.8) are precisely given. In the case of (1.3.8) the solution to the blow-up question is certainly not obvious due to the presence of the $-v^2$ and $-2uv$ terms.

1.3.2 Equations with Gradient Terms

The extension of blow-up results to the single equation

$$\frac{\partial u}{\partial t} = \Delta u + |u|^{p-1} u + \mathbf{a} \cdot \nabla (|u|^{q-1} u),$$

(1.3.9)

has been considered by several writers, e.g. Aguirre & Escobedo (1993), Bandle & Levine (1994), Friedman (1988), Friedman & Lacey (1988), Levine *et al.* (1989), Meier (1990).

Levine *et al.* (1989) consider the boundary initial value problem,

$$\frac{\partial u}{\partial t} + \epsilon \frac{\partial}{\partial x}\Big(g(u)\Big) = \frac{\partial^2 u}{\partial x^2} + f(u), \qquad x \in (0, L), \, t > 0,$$
$$u = 0, \qquad x = 0, L,$$
$$u(x, 0) = u_0(x).$$

(1.3.10)

The constant ϵ may be positive or negative and the functions f, g are typically given by $f(u) = u^p$, $g(u) = u^m$, with $m, p > 1$. They use a power eigenfunction argument to show that if $p > m$ and the initial data function u_0 satisfies the inequality

$$\int_0^L u_0 \phi_1^n \, dx > C_0,$$

for suitable constants n, C_0 then u blows up in finite time. This result is extended to the case where the functions f and g satisfy the following conditions:

$f(0) = 0$, $f(u)/u$ is strictly increasing on $\mathbf{R}^+$,

$$0 \leq \lim_{u \to 0} \frac{f(u)}{u} < \left(\frac{\pi}{L}\right)^2 < \lim_{u \to \infty} \frac{f(u)}{u},$$

$$g(0) = 0, g'(u) > 0 \text{ for } u > 0,$$

and

$$f(u) \geq C_1 u^p - C_2, \qquad u \geq 0,$$
$$g(u) \leq C_3 u^m + C_4, \qquad u \geq 0,$$

for constants C_i, with $p > m$. Essentially this shows that when the nonlinearity f behaves worse than g then blow-up can occur.

Meier (1990) and Bandle & Levine (1994) consider the problem

$$\frac{\partial u}{\partial t} = \Delta u + \mathbf{b} \cdot \nabla u + u^p, \qquad \text{in } D \times (0, T),$$
$$u = 0, \qquad \text{on } \partial D \times (0, T),$$
$$u(\mathbf{x}, 0) = u(\mathbf{x}) \geq 0, \qquad u(\mathbf{x}, t) \geq 0.$$

(1.3.11)

Here $p > 1$, $\mathbf{b}$ is an N-dimensional vector which may depend on $\mathbf{x}$; (Bandle & Levine (1994) allow $\mathbf{b}$ to also depend on u). The domain D is $\mathbf{R}^N$ or some other unbounded region.

Bandle & Levine (1994) show that for

$$f \geq u^p - \alpha u - \gamma,$$

α, γ constants, with

$$\mathbf{b} \cdot \nabla u = \mathbf{b}^{(1)}(\mathbf{x}) \cdot \nabla u + \nabla \cdot \mathbf{B}(u)$$

for $\mathbf{b}^{(1)}(\mathbf{x})$ suitable and

$$|\mathbf{B}(u)| \leq C_2(1 + u^q),$$

then with $q < p$ solutions to (1.3.11) need not be global (i.e. they can blow-up in finite time). They then give many detailed results when $\mathbf{b}$ is a function of $\mathbf{x}$ only and finally they consider the situation when $\mathbf{b}(u) \cdot \nabla u = \nabla \cdot \mathbf{B}(u)$ for $\mathbf{B}$ growing no faster than a polynomial. In both cases Bandle & Levine (1994) give a detailed proof of Fujita-like results establishing blow-up for a solution to (1.3.11).

Aguirre & Escobedo (1993) also give detailed Fujita-like results for (1.3.9) on the spatial domain $\mathbf{R}^N$. These writers also consider the problem (1.3.9) with initial data given at a point source.

Blow-up results for another equation with gradient terms of form

$$\frac{\partial u}{\partial t} = \Delta u + u^p - |\nabla u|^q, \tag{1.3.12}$$

are given by Chipot & Weissler (1989) and Fila (1991). Since the $-|\nabla u|^q$ term may act as a stabilizing ingredient the question as to whether u blows up in finite time or not is non-trivial. Chipot & Weissler (1989) treat (1.3.12) on a bounded domain $\Omega \subset \mathbf{R}^N$ with zero boundary data. They show blow-up can occur if

$$1 < q < \frac{2p}{p+1}.$$

It is noteworthy that Kawohl & Peletier (1989) also treat (1.3.12) with $q = 2$.

Blow-up and existence results are also established for the following boundary initial value problem by Junning (1993),

$$\frac{\partial u}{\partial t} = \frac{\partial}{\partial x_i}\left(|\nabla u|^{p-2}\frac{\partial u}{\partial x_i}\right) + f(\nabla u, u, x, t), \qquad \text{in } \Omega \times (0, T),$$

$$u(x, 0) = u_0(x), \qquad x \in \Omega,$$

$$u(x, t) = 0, \qquad x \in \Gamma,$$

where p is a constant, $p > 2$, and Ω is a bounded domain in $\mathbf{R}^N$ with boundary Γ. Of course, this partial differential equation may become degenerate in the sense that the diffusion coefficient may tend to zero as $|\nabla u| \to 0$.

It is also worth mentioning the study of Bricher (1994) who analyses the Cauchy problem and the boundary initial value problem for the partial differential equation

$$\frac{\partial u}{\partial t} = \nabla(k(u)\nabla u) + Q(u).$$

He is primarily interested in the case where

$$k(u) = 1 + \mu(u), \qquad Q(u) = u^p\left(1 + \nu(u)\right),$$

for suitable functions k and Q. Detailed estimates of blow - up behaviour are derived.

Budd & Galaktionov (1996) contains an interesting analysis for the partial differential equation

$$\frac{\partial u}{\partial t} = \frac{\partial}{\partial x}\left(\left|\frac{\partial u}{\partial x}\right|^\sigma \frac{\partial u}{\partial x}\right) + e^u,$$

for a non-negative constant σ. Much detail is given concerning the asymptotic behaviour for blowing up solutions.

1.3.3 Systems with Gradient Terms

With the motivation of studying non-linear fluid dynamical systems, such as those which arise in penetrative convection, see e.g. Straughan (1993), Ames & Straughan (1995) deal with the equations

$$
\begin{aligned}
Lu &= \alpha u^2 + \gamma v^2, \\
Lv &= \beta v^2 + \delta u^2,
\end{aligned}
\tag{1.3.13}
$$

where $\alpha, \beta, \gamma, \delta$ are positive constants and L is the operator

$$Lu = \frac{\partial u}{\partial t} + u\frac{\partial u}{\partial x} - \frac{\partial^2 u}{\partial x^2}.$$

For (1.3.12) on a bounded domain with zero boundary data it was shown that u, v remain bounded (in a suitable sense). Further, define the operators L_α, for $\alpha = 1, 2$, by

$$L_\alpha w = \frac{\partial w}{\partial t} + \epsilon_\alpha w\frac{\partial w}{\partial x} - \omega_\alpha \frac{\partial^2 w}{\partial x^2},$$

for $\epsilon_\alpha, \omega_\alpha$ positive constants. Then Ames & Straughan (1995) show that a global solution need not exist for the system

$$
\begin{aligned}
L_1 u &= \alpha u^{2+s} - \gamma v^2, \\
L_2 v &= \beta v^{2+s} - \delta u^2,
\end{aligned}
\tag{1.3.14}
$$

for $s > 0$. (Note that the last less than or equal to inequality in Ames & Straughan (1995) should be replaced by strictly less than.)

1.3.4 Equations with Gradient Terms and Non-Dirichlet Boundary Conditions

There are various interesting blow-up results for equations with gradient terms when the boundary conditions are ones of non-Dirichlet-type, e.g. Alikakos *et al.* (1989), Anderson (1991, 1993a, 1993b), Anderson & Deng (1995), Deng (1994) and Levine (1988).

The paper of Alikakos *et al.* (1989) is an interesting one which studies the boundary initial value problem

$$
\begin{aligned}
\frac{\partial u}{\partial t} &= \frac{\partial^2 u}{\partial x^2} + \frac{\partial}{\partial x}(u^q), \qquad 0 < x < 1,\ 0 < t < T, \\
\frac{\partial u}{\partial x} + u^q &= 0, \qquad x = 0, 1, \\
u(x,0) &= u_0(x),
\end{aligned}
\tag{1.3.15}
$$

where $q > 1$. They observe that equilibrium solutions, $\bar{u}$, exist which satisfy the equation

$$
\frac{\partial \bar{u}}{\partial x} + \bar{u}^q = 0,
$$

and so

$$
\bar{u} = \frac{1}{[(q-1)(x+k)]^{1/(q-1)}}, \qquad k \text{ constant.}
$$

The solution $\bar{u}$ which has its singularity at $x = 0$ is integrated over $(0,1)$ to yield a quantity called the critical mass, M_c, which may be calculated as

$$
M_c = \begin{cases} +\infty, & 1 < q \le 2, \\ \left(\frac{1}{q-1}\right)^{1/(q-1)}\left(\frac{q-1}{q-2}\right), & q > 2. \end{cases}
$$

Alikakos *et al.* (1989) show that if $q > 2$, $u_0 \ge 0$ and

$$
m = \int_0^1 u_0\,dx > M_c
$$

then $u(0,t)$ blows up in finite time. They further show that blow-up can only occur at the boundary $x = 0$. For the case when $q > 2$ and m is arbitrary then u may still blow-up in a finite time for certain u_0. The last result is particularly interesting since it shows that the critical mass condition is not the one which differentiates between blow-up or not, i.e. whether $\int_0^1 u_0(x)dx$ is greater than or less than $\int_0^1 \bar{u}\,dx$ is not the crucial factor.

Levine (1988) studies two non-standard boundary initial value problems for Burgers' equation. These are

$$\frac{\partial u}{\partial t} = \frac{\partial^2 u}{\partial x^2} + \epsilon u \frac{\partial u}{\partial x}, \qquad (x,t) \in (0,1) \times (0,\infty),$$

$$\frac{\partial u}{\partial x}(1,t) = au^p(1,t), \qquad t > 0, \qquad\qquad (1.3.16)$$

$$u(0,t) = 0, \qquad t > 0,$$

$$u(x,0) = u_0(x), \qquad x \in [0,1],$$

and

$$\frac{\partial u}{\partial t} = \frac{\partial^2 u}{\partial x^2} + \epsilon u \frac{\partial u}{\partial x}, \qquad (x,t) \in (0,1) \times (0,\infty),$$

$$\frac{\partial u}{\partial x}(0,t) = -au^p(0,t), \qquad t > 0, \qquad\qquad (1.3.17)$$

$$u(1,t) = 0, \qquad t > 0,$$

$$u(x,0) = u_0(x), \qquad x \in [0,1].$$

The numbers a, ϵ, p are positive constants. In addition, generalisations of the non-linear functions uu_x and u^p are examined.

It is not clear whether there is any physical system described by either of these boundary initial value problems. However, the study by Levine (1988) is a mathematical one. By using comparison arguments, the maximum principle, and the concavity method, Levine (1988) develops a detailed picture of the bifurcation diagram for systems (1.3.16) and (1.3.17), giving existence and non-existence results. The results of Levine (1988) are technical and we refer to the original paper for further details.

Anderson (1991) studies the boundary initial value problem

$$\frac{\partial u}{\partial t} = \nabla \cdot \left[\nabla \phi(\mathbf{x},t,u) + \mathbf{f}(\mathbf{x},t,u)\right] + h(\mathbf{x},t,u), \qquad \text{in } \Omega \times (0,T),$$

$$(\nabla \phi + \mathbf{f}) \cdot \mathbf{n} = g(\mathbf{x},t,u), \qquad \text{on } \Gamma \times (0,T), \qquad\qquad (1.3.18)$$

$$u(\mathbf{x},0) = u_0(\mathbf{x}), \qquad \mathbf{x} \in \Omega,$$

where $\Omega \subset \mathbf{R}^N$ is a bounded domain with boundary Γ and $\phi, \mathbf{f}, g$ are suitable non-linear functions. The blow-up result established in Anderson (1991) places various technical conditions on the functions and essential are the following,

$$\left|\frac{\partial g}{\partial u}(\mathbf{x},t,u)\right| \leq C \frac{\partial \phi}{\partial u}(\mathbf{x},t,u) \qquad \text{on } \Gamma,$$

$$f'(u), h(u), g - f \geq 0, \qquad \text{for } u \geq 0,$$

$$G(u) \leq \frac{1}{2}\phi(u)g(u), \qquad P(u) \leq \frac{1}{2}\phi(u)h(u),$$

and for $\kappa \in (0,1/2)$, Φ^κ is a convex function of u, where G, P, Φ are defined by

$$G(u) = \int_0^u \phi'(s)g(s)\,ds,$$

$$P(u) = \int_0^u \phi'(s)h(s)\,ds,$$

$$\Phi(u) = \int_0^u \phi(s)\,ds.$$

Then, for a precisely determined class of initial data, Anderson (1991) establishes u blows up in finite time.

The papers of Anderson (1993a, 1993b) are devoted to the problems

$$\frac{\partial u}{\partial t} = \frac{\partial^2}{\partial x^2}(u^m) + \frac{\epsilon}{n}\frac{\partial}{\partial x}(u^n), \qquad x \in (0,1), t > 0,$$

$$u(x,0) = u_0(x),$$

(1.3.19)

and *either*

$$u(0,t) = 0, \qquad \frac{\partial}{\partial x}(u^m)(1,t) = au^p(1,t),$$

(1.3.20)

or

$$u(1,t) = 0, \qquad \frac{\partial}{\partial x}(u^m)(0,t) = -au^p(0,t).$$

(1.3.21)

He studies how the solutions to (1.3.19) together with either of the boundary conditions (1.3.20) or (1.3.21) behave as functions of the parameters m, ϵ, n, p and a, and among other things he establishes conditions under which finite time blow-up of the solution is seen. The papers by Anderson (1993a, 1993b) generalise the work by Levine (1988) who dealt with (1.3.19) when $m = 1$ and $n = 2$, i.e. Burgers' equation with non-linear forcing at the boundary.

Deng (1994) investigates an analogous problem to that just mentioned, namely

$$\frac{\partial u}{\partial t} = \frac{\partial^2 u}{\partial x^2} + \frac{\partial}{\partial x}\big(f(u)\big),$$

but subject to the non-local boundary condition

$$\frac{\partial u}{\partial x}(0,t) + f\big(u(0,t)\big) = g\left(u(0,t), \int_0^1 u\,dx\right).$$

Detailed results are given for solution behaviour and among these conditions for blow-up are presented.

Anderson & Deng (1995) effectively consider the equation in (1.3.18) but subject to homogeneous boundary conditions

$$u = 0, \qquad x = 0, 1, t > 0.$$

Again, the goal is to give a detailed account of how solutions behave and among the behaviour found is that of blow-up in finite time.

1.3.5 Blow-Up of Derivatives

Fila & Lieberman (1994) takes a different line of attack and they examine the question of when the spatial derivative of the solution will blow up, *but the solution itself remains bounded.* The boundary initial value problem they examine is

$$\frac{\partial u}{\partial t} = \frac{\partial^2 u}{\partial x^2} + f\left(\frac{\partial u}{\partial x}\right), \qquad x \in (0, L),\ t > 0,$$

$$u = 0, \qquad x = 0, L, \quad t > 0, \tag{1.3.22}$$

$$u(x, 0) = u_0(x).$$

The function f is positive and $f'(v) \geq 0$ for v large enough, and in addition,

$$\limsup_{v \to \infty} \frac{f'(v)}{f(v)} < \infty,$$

$$\int^{\infty} \frac{v}{f(v)}\, dv < \infty, \tag{1.3.23}$$

and *either*

$$\int_{-\infty} \frac{dv}{f(v)} = \infty, \tag{1.3.24}$$

or

$$\int_{-\infty} \frac{dv}{f(v)} < \infty, \quad \text{together with} \quad \int_{-\infty} \frac{v}{f(v)}\, dv = -\infty. \tag{1.3.25}$$

(The notation of one limit in the integral indicates it is the behaviour at $\pm\infty$ which is important.) Fila & Lieberman (1994) demonstrate that the above conditions together with

$$\liminf_{v \to \infty} f(v) > 0$$

will guarantee that for L large enough $\partial u/\partial x$ blows up in finite time, regardless of the inital data u_0.

This result is of interest since Lieberman (1986) has shown that the conditions

$$\int^{\infty} \frac{v}{f(v)}\, dv = -\int_{-\infty} \frac{v}{f(v)}\, dv = \infty$$

guarantee $\partial u/\partial x$ and u do not blow-up. It is worth noting that Dlotko (1991) and Kutev (1991, 1992, 1994) have constructed examples to show that gradient blow-up may occur at the boundary when conditions like (1.3.23)–(1.3.25) are in force.

2. Analysis of a First-Order System

2.1 Conditional Decay of Solutions

In this chapter we investigate the properties of solutions to a system of partial differential equations motivated by the Navier-Stokes equations and ramifications of these which arise in various thermal convection contexts. In addition to studying blow-up (or non-existence) we also examine when the solution will remain bounded for all time, or even decay. The study of the system is of interest in its own right, and we have not seen this particular investigation elsewhere. The results reported here are an extension of sections 2.3–2.6 of Straughan (1992), and in places are an extension of Ames & Straughan (1995) and Levine et al. (1989).

Consider the convection–reaction–diffusion system of first order in time partial differential equations

$$\frac{\partial u}{\partial t} + \frac{\partial}{\partial x}(g_1(u)) = \frac{\partial^2 u}{\partial x^2} + f_1(u, v),$$
$$\frac{\partial v}{\partial t} + \frac{\partial}{\partial x}(g_2(v)) = \frac{\partial^2 v}{\partial x^2} + f_2(u, v),$$
(2.1.1)

defined on the domain $(x, t) \in (0, 1) \times (0, \infty)$. The functions g_1, g_2, f_1, f_2 will have various polynomial forms in this section, and in the subsequent sections which comprise this chapter.

In the current section we choose the functions g_1, g_2 such that

$$g_1(u) = \sum_{i=1}^{m} k_i u^i, \qquad g_2(v) = \sum_{i=1}^{n} \ell_i v^i, \qquad (2.1.2)$$

where k_i, ℓ_i are constants and $m, n \in \mathbb{N}$, but are otherwise arbitrary. Throughout this chapter we assume u, v are non-negative, although it is often possible to remove this restriction. The functions f_1, f_2 are such that

$$f_1(u, v) \le \alpha u^a + \gamma v^a, \qquad f_2(u, v) \le \beta v^a + \delta u^a, \qquad (2.1.3)$$

where $\alpha, \beta, \gamma, \delta$ are positive constants and $a \ge 1$ is otherwise arbitrary.

The boundary conditions satisfied by u, v are

$$u = 0, \quad v = 0, \qquad x = 0, 1, \qquad (2.1.4)$$

and the initial data are

$$u(x,0) = u_0(x), \qquad v(x,0) = v_0(x). \tag{2.1.5}$$

Let $(\cdot, \cdot)$ and $\|\cdot\|$ denote the inner product and norm on $L^2(0,1)$. We commence with an energy-like analysis and so multiply $(2.1.1)_1$ by u and integrate over $(0,1)$, and likewise multiply $(2.1.1)_2$ by v and integrate the outcome over the interval $(0,1)$. Note that by integration by parts, the forms $(2.1.2)$, and the boundary conditions,

$$\left(\frac{\partial}{\partial x}(g_1(u)), u \right) = 0, \qquad \left(\frac{\partial}{\partial x}(g_2(v)), v \right) = 0.$$

We find that

$$\frac{d}{dt} \frac{1}{2} \|u\|^2 = -\|u_x\|^2 + (u, f_1),$$

$$\leq -\|u_x\|^2 + \alpha \int_0^1 u^{a+1} dx + \gamma \int_0^1 uv^a dx, \tag{2.1.6}$$

and

$$\frac{d}{dt} \frac{1}{2} \|v\|^2 = -\|v_x\|^2 + (v, f_2),$$

$$\leq -\|v_x\|^2 + \beta \int_0^1 v^{a+1} dx + \delta \int_0^1 vu^a dx. \tag{2.1.7}$$

Notice that from Young's inequality one may show

$$v^a u \leq \left(\frac{a}{a+1} \right) v^{a+1} + \left(\frac{1}{a+1} \right) u^{a+1}, \tag{2.1.8}$$

and using this in $(2.1.6)$ and the equivalent inequality with the roles of u and v reversed in $(2.1.7)$, we obtain

$$\frac{d}{dt} \frac{1}{2} \|u\|^2 \leq -\|u_x\|^2 + \left(\alpha + \frac{\gamma}{a+1} \right) \int_0^1 u^{a+1} dx$$

$$+ \frac{\gamma a}{a+1} \int_0^1 v^{a+1} dx, \tag{2.1.9}$$

$$\frac{d}{dt} \frac{1}{2} \|v\|^2 \leq -\|v_x\|^2 + \left(\beta + \frac{\delta}{a+1} \right) \int_0^1 v^{a+1} dx$$

$$+ \frac{\delta a}{a+1} \int_0^1 u^{a+1} dx. \tag{2.1.10}$$

Addition of $(2.1.9)$ and $(2.1.10)$ produces

$$\frac{d}{dt}\frac{1}{2}(\|u\|^2 + \|v\|^2) \leq -(\|u_x\|^2 + \|v_x\|^2) + \left(\alpha + \frac{\gamma}{a+1} + \frac{\delta a}{a+1}\right)\int_0^1 u^{a+1}dx$$

$$+ \left(\beta + \frac{\delta}{a+1} + \frac{\gamma a}{a+1}\right)\int_0^1 v^{a+1}dx. \qquad (2.1.11)$$

Define the constant ζ by

$$\zeta = \max\left\{\alpha + \frac{\gamma}{a+1} + \frac{\delta a}{a+1}, \beta + \frac{\delta}{a+1} + \frac{\gamma a}{a+1}\right\}.$$

For an explicitly computable constant c we use the inequality

$$\int_0^1 u^{a+1}dx \leq c\|u_x\|^2\left(\int_0^1 u^2 dx\right)^{(a-1)/2}, \qquad (2.1.12)$$

see (2.1.18), and then from (2.1.11) we may derive

$$\frac{d}{dt}\frac{1}{2}(\|u\|^2 + \|v\|^2) \leq -(\|u_x\|^2 + \|v_x\|^2)$$

$$+ c\zeta(\|u_x\|^2\|u\|^{a-1} + \|v_x\|^2\|v\|^{a-1}),$$

$$\leq -(\|u_x\|^2 + \|v_x\|^2)$$

$$+ c\zeta(\|u_x\|^2 + \|v_x\|^2)(\|u\|^{a-1} + \|v\|^{a-1}). \quad (2.1.13)$$

Next, use the fact that

$$\|u\|^{a-1} + \|v\|^{a-1} \leq (\|u\|^2 + \|v\|^2)^{(a-1)/2},$$

to deduce from (2.1.13),

$$\frac{d}{dt}\frac{1}{2}(\|u\|^2 + \|v\|^2) \leq -(\|u_x\|^2 + \|v_x\|^2)\left[1 - c\zeta(\|u\|^2 + \|v\|^2)^{(a-1)/2}\right]. \quad (2.1.14)$$

If now

$$\left[\|u_0\|^2 + \|v_0\|^2\right]^{(a-1)/2} < \frac{1}{c\zeta}, \qquad (2.1.15)$$

then one may show $\|u(t)\|^2 + \|v(t)\|^2 \to 0$, exponentially fast, as $t \to \infty$, cf. Straughan (1992), p. 12. Thus, if the initial data is so small that (2.1.15) is satisfied, the solutions to (2.1.1)–(2.1.5) decay. Of course, condition (2.1.15) involves the sizes of the constants $\alpha, \beta, \gamma, \delta$ and a. Note that this result is true irrespective of the powers m, n, a in the expressions for g_α and f_α in (2.1.2) and (2.1.3).

To establish (2.1.12), note that by Hölder's inequality

$$\int_0^1 u^{a+1}dx \leq \left(\int_0^1 u^{2p}dx\right)^{1/p}\left(\int_0^1 u^{(a-1)q}dx\right)^{1/q},$$

for $p > 1$, arbitrary, $p^{-1} + q^{-1} = 1$. Thus put $2p = k$, then

$$\int_0^1 u^{a+1}dx \le \left(\int_0^1 u^k dx\right)^{2/k}\left(\int_0^1 u^{(a-1)q}dx\right)^{1/q},$$

$$\le c_1\|u_x\|^2\left(\int_0^1 u^{(a-1)q}dx\right)^{1/q}, \qquad (2.1.16)$$

where we have used the inequality

$$\|u\|_p \le c_1^{1/2}\|u_x\|, \qquad (2.1.17)$$

for $p > 1$ arbitrary, with $\|\cdot\|_p$ being the $L^p(0,1)$ norm. Now, pick $q = 2/(a-1)$ in (2.1.16) and so

$$\int_0^1 u^{a+1}dx \le c_1\|u_x\|^2\left(\int_0^1 u^2 dx\right)^{(a-1)/2}. \qquad (2.1.18)$$

An estimate for c_1 in (2.1.17) (not optimal) may be easily found. For example, note that for $x \in (0,L)$,

$$u^2(x) = 2\int_0^x uu'dx,$$

$$\le 2\left(\int_0^x u^2 dx\right)^{1/2}\left(\int_0^x (u')^2 dx\right)^{1/2},$$

using the Cauchy–Schwarz inequality. Hence, for $p > 1$,

$$u^p(x) \le 2^{p/2}\left(\int_0^L u^2 dx\right)^{p/4}\left(\int_0^L (u')^2 dx\right)^{p/4},$$

and so

$$\left(\int_0^L u^p dx\right)^{1/p} \le 2^{1/2}L^{1/p}\left(\int_0^L u^2 dx\right)^{1/4}\left(\int_0^L (u')^2 dx\right)^{1/4}. \qquad (2.1.19)$$

But, from Poincaré's inequality,

$$\frac{\pi^2}{L^2}\int_0^L u^2 dx \le \int_0^L (u')^2 dx,$$

and so using this in (2.1.19) leads to

$$\left(\int_0^L u^p dx\right)^{1/p} \le \frac{\sqrt{2}L^{(2+p)/2p}}{\sqrt{\pi}}\left(\int_0^L (u')^2 dx\right)^{1/2}.$$

The constant in this inequality represents an upper bound for the best value of $c_1^{1/2}$ in (2.1.17).

In this section we have shown that no matter how large the power of the forcing terms f_1, f_2 in (2.1.1), a solution may still decay. In the next section we show that (in a sense made precise there) if the power of the non-linearities f_1, f_2 does not exceed that of the convective non-linearities g_1 and g_2, then the solution (u, v) remains bounded for all time. This prevents a finite-time blow-up or global non-existence phenomenon.

2.2 Boundedness of Solutions

In this section we again investigate the properties of solutions to the system of partial differential equations (2.1.1) subject to the boundary and initial data (2.1.4), (2.1.5), under the assumption that $u, v \geq 0$. The non-linear functions f_α, g_α here, however, have the forms

$$g_1(u) = h_1 u^p, \qquad g_2(v) = h_2 v^p,$$
$$f_1 \leq \alpha u^p + \gamma v^p, \qquad f_2 \leq \beta v^p + \delta u^p, \tag{2.2.1}$$

for $p \geq 2$. The terms $h_1, h_2, \alpha, \beta, \gamma, \delta$ are positive constants.

We commence by introducing, for $k > 0$ to be chosen, the weight function

$$\mu(x) = e^{-kx}. \tag{2.2.2}$$

Then define F and G by

$$F(t) = \int_0^1 \mu u^2 dx, \qquad G(t) = \int_0^1 \mu v^2 dx. \tag{2.2.3}$$

By differentiation, substitution from the partial differential equation (2.1.1), use of (2.2.1), and integration by parts, we find

$$\frac{dF}{dt} = \frac{2}{(p+1)} \int_0^1 \mu_x u g_1 \, dx + 2 \int_0^1 \mu u f_1 dx$$
$$- 2 \int_0^1 \mu u_x^2 dx + \int_0^1 \mu_{xx} u^2 \, dx,$$
$$\leq -2 \left(\frac{kh_1}{p+1} - \alpha \right) \int_0^1 \mu u^{p+1} dx + 2\gamma \int_0^1 \mu u v^p dx$$
$$+ k^2 \int_0^1 \mu u^2 dx - 2 \int_0^1 \mu u_x^2 dx. \tag{2.2.4}$$

Likewise,

$$\frac{dG}{dt} \leq -2 \left(\frac{kh_2}{p+1} - \beta \right) \int_0^1 \mu v^{p+1} dx + 2\delta \int_0^1 \mu v u^p dx$$
$$+ k^2 \int_0^1 \mu v^2 dx - 2 \int_0^1 \mu v_x^2 dx. \tag{2.2.5}$$

By using Young's inequality we have

$$\int_0^1 \mu u v^p \, dx \leq \frac{1}{(p+1)} \int_0^1 \mu u^{p+1} \, dx + \frac{p}{(p+1)} \int_0^1 \mu v^{p+1} \, dx. \qquad (2.2.6)$$

We use inequality (2.2.6) in (2.2.4) and use the equivalent form to (2.2.6) with the roles of u and v reversed in (2.2.5). Then, add the resulting inequalities to see that

$$\begin{aligned}
\frac{d}{dt}(F+G) \leq & -2\left(\frac{kh_1}{p+1} - \alpha - \frac{\gamma}{p+1} - \frac{\delta p}{p+1}\right) \int_0^1 \mu u^{p+1} \, dx \\
& -2\left(\frac{kh_2}{p+1} - \beta - \frac{\delta}{p+1} - \frac{\gamma p}{p+1}\right) \int_0^1 \mu v^{p+1} \, dx \\
& -2 \int_0^1 \mu u_x^2 \, dx - 2 \int_0^1 \mu v_x^2 \, dx \\
& + k^2 \int_0^1 \mu u^2 \, dx + k^2 \int_0^1 \mu v^2 \, dx.
\end{aligned} \qquad (2.2.7)$$

We bound the third and fourth terms on the right of (2.2.7) by using the inequality, see Straughan (1992), p. 14,

$$-2 \int_0^1 \mu u_x^2 \, dx \leq -2\left(\frac{k^2}{4} + \pi^2\right) \int_0^1 \mu u^2 \, dx. \qquad (2.2.8)$$

Furthermore, we pick k so that

$$k = \max\left\{\frac{\alpha(p+1)}{h_1} + \frac{\gamma}{h_1} + \frac{\delta p}{h_1} + 1, \frac{\beta(p+1)}{h_2} + \frac{\delta}{h_2} + \frac{\gamma p}{h_2} + 1\right\}, \qquad (2.2.9)$$

and define γ_1 so that

$$\gamma_1 = \min\{h_1, h_2\}. \qquad (2.2.10)$$

Then, by using (2.2.8)–(2.2.10) in inequality (2.2.7) we may arrive at

$$\begin{aligned}
\frac{d}{dt}(F+G) \leq & -\frac{2\gamma_1}{p+1}\left[\int_0^1 \mu u^{p+1} \, dx + \int_0^1 \mu v^{p+1} \, dx\right] \\
& + \left(\frac{k^2}{2} - 2\pi^2\right)(F+G).
\end{aligned} \qquad (2.2.11)$$

Now, from Hölder's inequality observe that

$$\int_0^1 \mu u^2 \, dx \leq \left(\int_0^1 \mu u^{p+1} \, dx\right)^{2/(p+1)} \left(\int_0^1 \mu \, dx\right)^{(p-1)/(p+1)},$$

so that

$$\int_0^1 \mu u^{p+1} \, dx \geq \left(\frac{k}{1-e^{-k}}\right)^{(p-1)/2} \left(\int_0^1 \mu u^2 \, dx\right)^{(p+1)/2} \qquad (2.2.12)$$

Thus, from (2.2.11) we may show that

$$\frac{d}{dt}(F+G) \leq -\frac{2\gamma_1}{p+1}\left(\frac{k}{1-e^{-k}}\right)^{(p-1)/2}\left(F^{(p+1)/2}+G^{(p+1)/2}\right)$$
$$+\left(\frac{k^2}{2}-2\pi^2\right)(F+G). \tag{2.2.13}$$

Next use the inequality,

$$2^{(p-1)/2}\left(F^{(p+1)/2}+G^{(p+1)/2}\right) \geq (F+G)^{(p+1)/2}, \qquad p \geq 2,$$

to find from (2.2.13) that

$$\frac{d}{dt}(F+G) \leq -\zeta_1(F+G)^{(p+1)/2}+\zeta_2(F+G), \tag{2.2.14}$$

where the coefficients ζ_1, ζ_2 are given by

$$\zeta_1 = \frac{2^{(3-p)/2}\gamma_1}{(1+p)}\left(\frac{k}{1-e^{-k}}\right)^{(p-1)/2},$$

$$\zeta_2 = \frac{1}{2}k^2 - 2\pi^2.$$

Inequality (2.2.14) is of Bernoulli type and may be integrated to see that

$$\left[F(t)+G(t)\right]^{(p-1)/2}$$
$$\leq \frac{1}{e^{-\omega t}\left[F(0)+G(0)\right]^{(1-p)/2}+(\xi/\omega)(1-e^{-\omega t})}, \tag{2.2.15}$$

where the coefficients ξ, ω have been defined by

$$\xi = \frac{1}{2}\zeta_1(p-1), \qquad \omega = \frac{1}{2}\zeta_2(p-1).$$

Note that

$$F(t)+G(t) \geq e^{-k}\left(\|u(t)\|^2+\|v(t)\|^2\right), \tag{2.2.16}$$

and as $t \to \infty$, the right hand side of (2.2.15) approaches the limit

$$\frac{\omega}{\xi} = \frac{(k^2-4\pi^2)(1+p)(1-e^{-k})^{(p-1)/2}}{2^{(5-p)/2}\gamma_1 k^{(p-1)/2}}. \tag{2.2.17}$$

Thus, from (2.2.15)–(2.2.17), we find for $t \to \infty$, the asymptotic behaviour,

$$\|u(t)\|^2+\|v(t)\|^2 \leq \frac{e^k\omega}{\xi}.$$

While this is not as clear as the single equation case, cf. Straughan (1992), p. 15, it clearly shows u and v remain bounded (in the sense implied by (2.2.15) and (2.2.16)) for all $t > 0$. However, u and v may have very large gradients in x, and indeed in some cases do. This is evident in the numerical results displayed in Sect. 2.5.

2.3 Unconditional Decay of Solutions

We now establish a result for a solution to the system of partial differential equations (2.1.1) which shows that provided the power of the non-linearities f_1, f_2 does not exceed that of g_1, g_2, and the coefficients in the non-linearities f_1, f_2 are not too large, then the solution will decay exponentially for all initial data.

We essentially are interested in the functions f_α, g_α as defined in (2.2.1) in the last section. However, we here restrict attention to the case of $p = 2$. Observe that no unconditional decay result was given in Ames & Straughan (1995). Therefore, let u, v solve the system of partial differential equations

$$\frac{\partial u}{\partial t} + u\frac{\partial u}{\partial x} = \frac{\partial^2 u}{\partial x^2} + \alpha u^2 + \gamma v^2,$$
$$\frac{\partial v}{\partial t} + v\frac{\partial v}{\partial x} = \frac{\partial^2 v}{\partial x^2} + \beta v^2 + \delta u^2,$$

$$(2.3.1)$$

subject to the homogeneous boundary conditions (2.1.4) and initial data (2.1.5). We again assume $u, v \geq 0$, although this is not always necessary.

For a number $\varsigma > 0$ to be chosen, we now define the functions $F(t)$ and $G(t)$ by

$$F(t) = \int_0^1 \mu u^{1+\varsigma} dx, \qquad G(t) = \int_0^1 \mu v^{1+\varsigma} dx, \qquad (2.3.2)$$

where $\mu(x)$ is the weight function introduced in (2.2.2). We differentiate F and integrate by parts, using (2.3.1), to find

$$\frac{dF}{dt} = (1+\varsigma)\int_0^1 \mu u^\varsigma \left(-u\frac{\partial u}{\partial x} + \frac{\partial^2 u}{\partial x^2} + \alpha u^2 + \gamma v^2\right) dx,$$

$$= \left(\frac{1+\varsigma}{2+\varsigma}\right)\int_0^1 \mu_x u^{2+\varsigma} dx - (1+\varsigma)\int_0^1 \mu_x u^\varsigma u_x dx$$

$$- \varsigma(1+\varsigma)\int_0^1 \mu u^{\varsigma-1}u_x^2 dx + \alpha(1+\varsigma)\int_0^1 \mu u^{2+\varsigma} dx$$

$$+ \gamma(1+\varsigma)\int_0^1 \mu u^\varsigma v^2 dx. \qquad (2.3.3)$$

Define r and s by
$$r = u^{(1+\varsigma)/2}, \qquad s = v^{(1+\varsigma)/2}. \qquad (2.3.4)$$

The after some further rearrangement in (2.3.3) we may derive

$$\frac{dF}{dt} = \left(\frac{1+\zeta}{2+\zeta}\right) \int_0^1 \mu_x u^{2+\zeta} dx + \int_0^1 \mu_{xx} u^{1+\zeta} dx$$

$$- \frac{4\zeta}{(1+\zeta)} \int_0^1 \mu r_x^2 \, dx + \alpha(1+\zeta) \int_0^1 \mu u^{2+\zeta} dx$$

$$+ \gamma(1+\zeta) \int_0^1 \mu u^\zeta v^2 \, dx. \tag{2.3.5}$$

By a similar calculation,

$$\frac{dG}{dt} = \left(\frac{1+\zeta}{2+\zeta}\right) \int_0^1 \mu_x v^{2+\zeta} dx + \int_0^1 \mu_{xx} v^{1+\zeta} dx$$

$$- \frac{4\zeta}{(1+\zeta)} \int_0^1 \mu s_x^2 \, dx + \beta(1+\zeta) \int_0^1 \mu v^{2+\zeta} dx$$

$$+ \delta(1+\zeta) \int_0^1 \mu v^\zeta u^2 \, dx. \tag{2.3.6}$$

We next use the inequality

$$\int_0^1 \mu_{xx} u^{1+\zeta} dx \le k^2 \tilde{\lambda}_1^{-1} \int_0^1 \mu r_x^2 dx, \tag{2.3.7}$$

where

$$\tilde{\lambda}_1^{-1} = \max_{H_0^1(0,1)} \frac{\int_0^1 \mu w^2 dx}{\int_0^1 \mu w_x^2 dx} = \pi^2 + \frac{k^2}{4},$$

cf. Straughan (1992), p. 16. Inequality (2.3.7) is employed in (2.3.5) and the analogous expression involving v and s is employed in (2.3.6) to obtain

$$\frac{dF}{dt} \le (1+\zeta)\left(\alpha - \frac{k}{2+\zeta}\right) \int_0^1 \mu u^{2+\zeta} dx$$

$$- \frac{4\zeta}{(1+\zeta)}\left[1 - \frac{k^2(1+\zeta)}{4\zeta\tilde{\lambda}_1}\right] \int_0^1 \mu r_x^2 \, dx$$

$$+ \gamma(1+\zeta) \int_0^1 \mu u^\zeta v^2 \, dx, \tag{2.3.8}$$

and

$$\frac{dG}{dt} \le (1+\zeta)\left(\beta - \frac{k}{2+\zeta}\right) \int_0^1 \mu v^{2+\zeta} dx$$

$$- \frac{4\zeta}{(1+\zeta)}\left[1 - \frac{k^2(1+\zeta)}{4\zeta\tilde{\lambda}_1}\right] \int_0^1 \mu s_x^2 \, dx$$

$$+ \delta(1+\zeta) \int_0^1 \mu v^\zeta u^2 \, dx. \tag{2.3.9}$$

From Young's inequality we derive

$$\int_0^1 \mu u^\varsigma v^2 dx \leq \left(\frac{\varsigma}{2+\varsigma}\right) \int_0^1 \mu u^{2+\varsigma} dx$$
$$+ \left(\frac{2}{2+\varsigma}\right) \int_0^1 \mu v^{2+\varsigma} dx, \qquad (2.3.10)$$

with an equivalent inequality with the roles of u and v reversed. These are used in (2.3.8), (2.3.9) and the resulting inequalities are added to find that

$$\frac{d}{dt}(F+G) \leq (1+\varsigma)\left[\alpha + \frac{\gamma\varsigma}{(2+\varsigma)} + \frac{2\delta(1+\varsigma)}{(2+\varsigma)} - \frac{k}{(2+\varsigma)}\right]\int_0^1 \mu u^{2+\varsigma} dx$$
$$+ (1+\varsigma)\left[\beta + \frac{\delta\varsigma}{(2+\varsigma)} + \frac{2\gamma(1+\varsigma)}{(2+\varsigma)} - \frac{k}{(2+\varsigma)}\right]\int_0^1 \mu v^{2+\varsigma} dx$$
$$- \frac{4\varsigma}{(1+\varsigma)}\left[1 - \frac{k^2(1+\varsigma)}{4\varsigma\tilde{\lambda}_1}\right]\left(\int_0^1 \mu r_x^2 dx + \int_0^1 \mu s_x^2 dx\right). \quad (2.3.11)$$

We now choose k to ensure the last term in (2.3.11) is negative and also in such a way that the first two terms on the right of (2.3.11) are negative. Thus we select

$$k = \max\{\alpha(2+\varsigma) + \gamma\varsigma + 2\delta(1+\varsigma), \beta(2+\varsigma) + \delta\varsigma + 2\gamma(1+\varsigma)\}, \quad (2.3.12)$$

and

$$k^2(1+\varsigma) < 4\varsigma\tilde{\lambda}_1.$$

The last inequality reduces to

$$k^2 < 4\varsigma\pi^2. \qquad (2.3.13)$$

Provided $\alpha, \beta, \gamma, \delta$ are not too large we can select k such that (2.3.12) and (2.3.13) hold. Then, recollecting

$$F = \int_0^1 \mu r^2 dx, \qquad G = \int_0^1 \mu s^2 dx,$$

we see that from (2.3.11) we may obtain

$$\frac{d}{dt}(F+G) \leq -a\left(\int_0^1 \mu r_x^2 dx + \int_0^1 \mu s_x^2 dx\right),$$

for $a > 0$ and exponential decay follows easily from this inequality with the aid of Poincaré's inequality.

To make the results of this section sharp we need to be able to simultaneously satisfy (2.3.12) and (2.3.13) in an optimal way. We show how this may be achieved in some special cases.

2.3.1 Special Cases

Select $\alpha = \beta, \delta = \gamma$. (In this case we can modify the above proof to have any u, v, not just non-negative ones.) In this situation we pick

$$k = \alpha(2 + \zeta) + \gamma(3\zeta + 2).$$

Let $\gamma = \mu\alpha$ (for $\mu > 0$ a constant, not the weight function defined earlier). Then inequality (2.3.13) is

$$\alpha^2 < f(\zeta) = \frac{4\pi^2\zeta}{\left[2(1 + \mu) + (1 + 3\mu)\zeta\right]^2}.$$

The function $f(\zeta)$ has a maximum where

$$\zeta_m = \frac{2(1 + \mu)}{(1 + 3\mu)}.$$

Thus, if

$$\alpha^2 < \frac{\pi^2}{2(1 + \mu)(1 + 3\mu)},$$

all solutions to (2.3.1), (2.1.4), (2.1.5) decay, regardless of the size of u_0, v_0.

Note that when $\gamma = 0$, the above requires $\zeta = 2$ and

$$\alpha^2 < \frac{1}{2}\pi^2.$$

This is thus in agreement with the result for the single equation case given in Straughan (1992), p. 16.

Select now $\alpha = \beta = 0, \delta = \gamma$. We here find we require

$$\gamma^2 < g(\zeta) = \frac{4\zeta\pi^2}{(3\zeta + 2)^2}.$$

Thus, for g to have a maximum we pick $\zeta = 2/3$. Then the restriction guaranteeing unconditional decay of the solution u, v is

$$\gamma^2 < \frac{\pi^2}{6}.$$

2.4 Global Non-existence of Solutions

The object of this section is to investigate the behaviour of a solution to the system of partial differential equations

$$\frac{\partial u}{\partial t} + h_1 \frac{\partial}{\partial x} u^p = \frac{\partial^2 u}{\partial x^2} + \alpha u^{p+\delta} - \gamma v^p,$$

$$\frac{\partial v}{\partial t} + h_2 \frac{\partial}{\partial x} v^p = \frac{\partial^2 v}{\partial x^2} + \beta v^{p+\delta} - \theta u^p,$$

(2.4.1)

subject to the boundary conditions

$$u(0,t) = u(1,t) = v(0,t) = v(1,t) = 0,$$

(2.4.2)

and satisfying the initial conditions

$$u(x,0) = u_0(x), \qquad v(x,0) = v_0(x).$$

(2.4.3)

The coefficients $h_1, h_2, \alpha, \beta, \gamma, \theta$ are positive constants, $p \geq 2$, and $\delta > 0$. We show effectively that when the polynomial non-linearities f_1, f_2 in (2.1.1) are of higher order than the convective non-linearities g_1, g_2 then a solution u, v need not exist for all time. The expected behaviour is that of blow-up in finite time as experienced in the single equation case in Levine *et al.* (1989).

In this section we use the multi-eigenfunction argument and so define the functions $F(t)$ and $R(t)$ by

$$F(t) = \int_0^1 \phi^n u(x,t)\, dx, \qquad R(t) = \int_0^1 \phi^n v(x,t)\, dx,$$

(2.4.4)

where $\phi(x)$ is the first eigenfunction in the membrane problem for the spatial domain $\{x | x \in (0,1)\}$, i.e. $\phi = \sin \pi x$, and n is a positive number to be selected.

By differentiation, use of the partial differential equations (2.4.1), and integration by parts we find

$$\frac{dF}{dt} = \int_0^1 \phi^n \left(-h_1 \frac{\partial}{\partial x} u^p + \frac{\partial^2 u}{\partial x^2} + \alpha u^{p+\delta} - \gamma v^p \right) dx,$$

$$= n(n-1) \int_0^1 \phi^{n-2} \phi_x^2 u\, dx - n\pi^2 F + h_1 n \int_0^1 \phi^{n-1} \phi_x u^p\, dx$$

$$+ \alpha \int_0^1 \phi^n u^{p+\delta}\, dx - \gamma \int_0^1 \phi^n v^p\, dx.$$

(2.4.5)

We have not found a useful role for the first term on the right of (2.4.5) but since it is non-negative we discard it. We then choose

$$n = 1 + \frac{p}{\delta}$$

(2.4.6)

and use Hölder's inequality to see that

$$\int_0^1 \phi^{np/(p+\delta)} u^p \, dx \le \left(\int_0^1 1 \, dx \right)^{\delta/(p+\delta)} \left(\int_0^1 \phi^n u^{p+\delta} \, dx \right)^{p/(p+\delta)},$$

or by rearrangement,

$$\int_0^1 \phi^{1+p/\delta} u^{p+\delta} \, dx \ge \left(\int_0^1 \phi^{p/\delta} u^p \, dx \right)^{(p+\delta)/p} \tag{2.4.7}$$

Further use of Hölder's inequality shows

$$\int_0^1 u \phi^{1+p/\delta} \, dx \le \left(\int_0^1 \phi^{[1+(p-1)/\delta]\{p/(p-1)\}} \, dx \right)^{(p-1)/p} \left(\int_0^1 u^p \phi^{p/\delta} \, dx \right)^{1/p},$$

so that

$$\int_0^1 u^p \phi^{p/\delta} \, dx \ge \left(\int_0^1 u \phi^{1+p/\delta} \, dx \right)^p. \tag{2.4.8}$$

Next, define the functions G and S by

$$G(t) = \int_0^1 \phi^{p/\delta} u^p \, dx, \qquad S(t) = \int_0^1 \phi^{p/\delta} v^p \, dx, \tag{2.4.9}$$

and observe that thanks to (2.4.8),

$$G^{1/p} \ge F, \qquad S^{1/p} \ge R. \tag{2.4.10}$$

Hence, using (2.4.6)–(2.4.9) in (2.4.5) we may establish that

$$\frac{dF}{dt} \ge - n\pi^2 G^{1/p} - h_1 n\pi G$$
$$+ \alpha G^{(p+\delta)/p} - \gamma \int_0^1 \phi^n v^p \, dx. \tag{2.4.11}$$

Since $\phi \le 1$,

$$\int_0^1 \phi^n v^p \, dx \le S,$$

and so from (2.4.11) we may derive

$$\frac{dF}{dt} \ge -n\pi^2 G^{1/p} - h_1 n\pi G + \alpha G^{(p+\delta)/p} - \gamma S. \tag{2.4.12}$$

An analogous calculation involving R and the partial differential equation for v leads to

$$\frac{dR}{dt} \ge -n\pi^2 S^{1/p} - h_2 n\pi S + \beta S^{(p+\delta)/p} - \theta G. \tag{2.4.13}$$

The next step is to add inequalities (2.4.12) and (2.4.13) and define the constants c_1 and c_2 by

$$c_1 = \min\{\alpha, \beta\}, \qquad c_2 = \max\{n\pi h_1 + \theta, n\pi h_2 + \gamma\}.$$

This leads to

$$\frac{d}{dt}(F + R) \geq -n\pi^2 \left(G^{1/p} + S^{1/p}\right) - c_2(S + G)$$
$$+ c_1\left(G^{(p+\delta)/p} + S^{(p+\delta)/p}\right). \tag{2.4.14}$$

Use is now made of the following two inequalities,

$$G^{1/p} + S^{1/p} \leq 2^{1-1/p}(G + S)^{1/p}, \tag{2.4.15}$$

and

$$G^{(p+\delta)/p} + S^{(p+\delta)/p} \geq \frac{1}{2^{\delta/p}}(G + S)^{1+\delta/p}. \tag{2.4.16}$$

Upon employing these inequalities in (2.4.14) we arrive at

$$\Theta' \geq -n\pi^2 \Xi^{1/p} - 2^{1-p}c_2\Xi + 2^{1-p-\delta}c_1\Xi^{1+\delta/p}, \tag{2.4.17}$$

where the functions Θ, Ξ are given by

$$\Theta(t) = F(t) + R(t),$$
$$\Xi(t) = 2^{p-1}\left[G(t) + S(t)\right]. \tag{2.4.18}$$

Define now the function Q by

$$Q(Y) = c_1 2^{1-p-\delta} Y^{p+\delta} - n\pi^2 Y - c_2 2^{1-p} Y^2. \tag{2.4.19}$$

Suppose the initial data are such that

$$Q(\Theta(0)) > 0. \tag{2.4.20}$$

Then by continuity there exists a $t_1 > 0$ such that

$$Q(\Theta(t)) > 0, \qquad t \in [0, t_1).$$

By calculation of Q' we see that

$$\Theta Q'(\Theta) \geq Q(\Theta) > 0, \qquad t \in [0, t_1).$$

Hence Q is increasing in Θ for $t \in [0, t_1)$. Since

$$\Xi \geq \Theta^p, \tag{2.4.21}$$

see (2.4.25), then

$$Q(\Xi^{1/p}) \geq Q(\Theta) > 0, \qquad t \in [0, t_1). \tag{2.4.22}$$

Thus, from inequality (2.4.17),

$$\Theta' \geq Q(\Xi^{1/p}),$$
$$\geq Q(\Theta), \tag{2.4.23}$$

for $t \in [0, t_1)$, by (2.4.22).

Hence, since Q is increasing on $[0, t_1)$, $Q(\Theta(t_1)) \neq 0$, and so t_1 must either be infinite or the limit of existence of the solution. Separate variables in (2.4.23) to obtain

$$t_1 \leq \int_{\Theta(0)}^{\Theta(t_1)} \frac{d\Theta}{Q(\Theta)} \leq \int_{\Theta(0)}^{\infty} \frac{d\Theta}{Q(\Theta)} < \infty . \tag{2.4.24}$$

This leads to a contradiction if t_1 is infinite, and so we deduce the solution ceases to exist globally in a finite time.

This establishes the global non-existence result alluded to earlier. It remains to prove inequality (2.4.21). To this end observe that from (2.4.10),

$$G^{1/p} + S^{1/p} \geq F + R,$$

and so using inequality (2.4.15),

$$\begin{aligned} (G + S)^{1/p} &\geq 2^{(1-p)/p} (G^{1/p} + S^{1/p}), \\ &\geq 2^{(1-p)/p} (F + R). \end{aligned}$$

Therefore,

$$2^{p-1}(G + S) \geq (F + R)^p . \tag{2.4.25}$$

This inequality is the same as (2.4.21).

2.5 Numerical Results by Finite Elements

2.5.1 Solution Structure with Linear and Quadratic Right-Hand Sides

In Straughan *et al.* (1987) numerical solutions by a finite element method are obtained for the local and non-local partial differential equations

$$\begin{aligned} \frac{\partial u}{\partial t} + u \frac{\partial u}{\partial x} - \frac{1}{R} \frac{\partial^2 u}{\partial x^2} &= u - Ru\|u\|^2, \\ \frac{\partial u}{\partial t} + u \frac{\partial u}{\partial x} - \frac{1}{R} \frac{\partial^2 u}{\partial x^2} &= u, \\ \frac{\partial u}{\partial t} + u \frac{\partial u}{\partial x} - \frac{1}{R} \frac{\partial^2 u}{\partial x^2} &= u + Ru\|u\|, \end{aligned} \tag{2.5.1}$$

with emphasis being on computing steady states by solving the time dependent problems. In Straughan (1992) further results are given for $(2.5.1)_2$ and the work is extended to the partial differential equations

$$\frac{\partial u}{\partial t} + u\frac{\partial u}{\partial x} - \frac{1}{R}\frac{\partial^2 u}{\partial x^2} = u + Ru\|u\|^{\epsilon},$$

$$\frac{\partial u}{\partial t} + u\frac{\partial u}{\partial x} - \frac{1}{R}\frac{\partial^2 u}{\partial x^2} = u + Ru^2.$$

(2.5.2)

For all systems in (2.5.1), (2.5.2), the spatial domain is $x \in (0,1)$ and the boundary conditions are that $u = 0$ at $x = 0, 1$.

One feature which is strongly observed in the numerically computed steady state solutions to $(2.5.1)_2$ and $(2.5.2)_2$ are the structures. There may be two steady state solutions (apart from the zero one), one positive while the other is a reflection in the x-axis and about the line $x = 1/2$. We discuss the positive one here. There is a strong boundary layer near $x = 1$, where diffusion is clearly important, i.e. the effect of the $R^{-1}\partial^2 u/\partial x^2$ term is felt. However, across the rest of the spatial layer the solution appears to be diffusionless. This may be explained by the following heuristic argument which was shown to me by Professor Keith Miller of the University of California at Berkeley.

For equation $(2.5.1)_2$ without diffusion, i.e. without the $R^{-1}\partial^2 u/\partial x^2$ term, the solution u satisfies the partial differential equation

$$\frac{\partial u}{\partial t} + u\frac{\partial u}{\partial x} = u.$$

By characteristics,

$$\frac{du}{dt} = u, \qquad \text{on} \qquad \frac{dx}{dt} = u,$$

so

$$\frac{du}{dx} = 1.$$

This is clearly seen in the computations of Straughan *et al.* (1987). The steady state solutions to $(2.5.1)_2$ are linear across the layer with gradient 1, and a boundary layer is evident near $x = 1$, this boundary layer being steeper as R increases.

For equation $(2.5.2)_2$ the steady state has exponential growth across the layer. This may be seen by considering the partial differential equation

$$\frac{\partial u}{\partial t} + u\frac{\partial u}{\partial x} - \frac{1}{R}\frac{\partial^2 u}{\partial x^2} = u^2.$$

Without the diffusion term, an analogous argument to that given above, involving characteristics shows

$$\frac{du}{dx} = u.$$

Thus u looks like e^x across the layer. Again, the numerically computed steady solutions to $(2.5.2)_2$ confirm that u increases exponentially across the layer from $x = 0$, with a steep boundary layer near $x = 1$.

There are many numerical schemes for solving systems of partial differential equations with convection terms. For example, in their book, Bellomo & Preziosi (1995), p. 287, describe a Chebyshev collocation (numerical) technique for computing a solution to the convection–diffusion equation

$$\frac{\partial u}{\partial t} + c(x,t)\frac{\partial u}{\partial x} = \nu\frac{\partial^2 u}{\partial x^2} + F(x,t), \qquad x \in (a,b),\ t > 0,$$

for given initial data, and for the boundary conditions,

$$\alpha\frac{\partial u}{\partial x}(a,t) + \beta u(a,t) = f_1(t),$$

$$\gamma\frac{\partial u}{\partial x}(b,t) + \delta u(b,t) = f_2(t).$$

Bellomo & Preziosi (1995), p. 305, also give an explicit finite difference routine for solving the equation

$$\frac{\partial u}{\partial t} + c\frac{\partial u}{\partial x} = \left(\frac{\kappa}{1+\epsilon u}\right)\frac{\partial^2 u}{\partial x^2} - \left(\frac{\epsilon\kappa}{(1+\epsilon u)^2}\right)\left(\frac{\partial u}{\partial x}\right)^2,$$

for zero boundary conditions. In this section we study the solution behaviour to a system of form (2.4.1) by a finite element method.

We have seen that at least one of the powers in the α and β forcing terms on the right hand side of system (2.4.1) must be greater than p ($\delta > 0$) for the possibility of finite time blow-up to arise. In this section, therefore, we report numerical computations of solution structure when the right hand side is a combination of linear and quadratic terms while the convective non-linearity is for $p = 2$. There are essentially eight representative cases to consider. These are comprised of the following boundary initial value problems for the systems of partial differential equations given below:

System A

$$\frac{\partial u}{\partial t} + u\frac{\partial u}{\partial x} - \frac{1}{R}\frac{\partial^2 u}{\partial x^2} = \alpha u + \gamma v, \qquad x \in (0,1), t > 0,$$

$$\frac{\partial v}{\partial t} + v\frac{\partial v}{\partial x} - \frac{1}{R}\frac{\partial^2 v}{\partial x^2} = \delta u + \beta v, \qquad x \in (0,1), t > 0, \qquad (2.5.3)$$

$$u(0,t) = u(1,t) = v(0,t) = v(1,t) = 0,$$

$$u(x,0) = u_0(x), \quad v(x,0) = v_0(x);$$

System B

$$\frac{\partial u}{\partial t} + u\frac{\partial u}{\partial x} - \frac{1}{R}\frac{\partial^2 u}{\partial x^2} = \alpha u^2 + \gamma v, \qquad x \in (0,1), t > 0,$$

$$\frac{\partial v}{\partial t} + v\frac{\partial v}{\partial x} - \frac{1}{R}\frac{\partial^2 v}{\partial x^2} = \delta u + \beta v, \qquad x \in (0,1), t > 0, \qquad (2.5.4)$$

$$u(0,t) = u(1,t) = v(0,t) = v(1,t) = 0,$$

$$u(x,0) = u_0(x), \quad v(x,0) = v_0(x);$$

System C

$$\frac{\partial u}{\partial t} + u\frac{\partial u}{\partial x} - \frac{1}{R}\frac{\partial^2 u}{\partial x^2} = \alpha u^2 + \gamma v^2, \qquad x \in (0,1), t > 0,$$

$$\frac{\partial v}{\partial t} + v\frac{\partial v}{\partial x} - \frac{1}{R}\frac{\partial^2 v}{\partial x^2} = \delta u + \beta v, \qquad x \in (0,1), t > 0, \qquad (2.5.5)$$

$$u(0,t) = u(1,t) = v(0,t) = v(1,t) = 0,$$

$$u(x,0) = u_0(x), \quad v(x,0) = v_0(x);$$

System D

$$\frac{\partial u}{\partial t} + u\frac{\partial u}{\partial x} - \frac{1}{R}\frac{\partial^2 u}{\partial x^2} = \alpha u + \gamma v^2, \qquad x \in (0,1), t > 0,$$

$$\frac{\partial v}{\partial t} + v\frac{\partial v}{\partial x} - \frac{1}{R}\frac{\partial^2 v}{\partial x^2} = \delta u + \beta v, \qquad x \in (0,1), t > 0, \qquad (2.5.6)$$

$$u(0,t) = u(1,t) = v(0,t) = v(1,t) = 0,$$

$$u(x,0) = u_0(x), \quad v(x,0) = v_0(x);$$

System E

$$\frac{\partial u}{\partial t} + u\frac{\partial u}{\partial x} - \frac{1}{R}\frac{\partial^2 u}{\partial x^2} = \alpha u^2 + \gamma v, \qquad x \in (0,1), t > 0,$$

$$\frac{\partial v}{\partial t} + v\frac{\partial v}{\partial x} - \frac{1}{R}\frac{\partial^2 v}{\partial x^2} = \delta u^2 + \beta v, \qquad x \in (0,1), t > 0, \qquad (2.5.7)$$

$$u(0,t) = u(1,t) = v(0,t) = v(1,t) = 0,$$

$$u(x,0) = u_0(x), \quad v(x,0) = v_0(x);$$

System F

$$\frac{\partial u}{\partial t} + u\frac{\partial u}{\partial x} - \frac{1}{R}\frac{\partial^2 u}{\partial x^2} = \alpha u^2 + \gamma v, \qquad x \in (0,1), t > 0,$$

$$\frac{\partial v}{\partial t} + v\frac{\partial v}{\partial x} - \frac{1}{R}\frac{\partial^2 v}{\partial x^2} = \delta u + \beta v^2, \qquad x \in (0,1), t > 0, \qquad (2.5.8)$$

$$u(0,t) = u(1,t) = v(0,t) = v(1,t) = 0,$$

$$u(x,0) = u_0(x), \quad v(x,0) = v_0(x);$$

System G

$$\frac{\partial u}{\partial t} + u\frac{\partial u}{\partial x} - \frac{1}{R}\frac{\partial^2 u}{\partial x^2} = \alpha u + \gamma v^2, \qquad x \in (0,1), t > 0,$$

$$\frac{\partial v}{\partial t} + v\frac{\partial v}{\partial x} - \frac{1}{R}\frac{\partial^2 v}{\partial x^2} = \delta u^2 + \beta v, \qquad x \in (0,1), t > 0, \qquad (2.5.9)$$

$$u(0,t) = u(1,t) = v(0,t) = v(1,t) = 0,$$

$$u(x,0) = u_0(x), \quad v(x,0) = v_0(x);$$

System H

$$\frac{\partial u}{\partial t} + u\frac{\partial u}{\partial x} - \frac{1}{R}\frac{\partial^2 u}{\partial x^2} = \alpha u^2 + \gamma v^2, \qquad x \in (0,1), t > 0,$$

$$\frac{\partial v}{\partial t} + v\frac{\partial v}{\partial x} - \frac{1}{R}\frac{\partial^2 v}{\partial x^2} = \delta u^2 + \beta v^2, \qquad x \in (0,1), t > 0, \qquad (2.5.10)$$

$$u(0,t) = u(1,t) = v(0,t) = v(1,t) = 0,$$

$$u(x,0) = u_0(x), \quad v(x,0) = v_0(x).$$

The solution behaviour to these systems is investigated numerically with a semi implicit finite element method using piecewise linear basis functions, dividing (0,1) into 1000 intervals, although higher spatial resolution was used as a check. Newton's method is used on the non-linear discrete system that arises and the 1998×1998 banded tridiagonal linear system of form $Ax = b$ that arises at each Newton update is solved using the preconditioned conjugate gradient squared technique of Sonneveld (1989). As a check, symmetric successive over relaxation (SSOR) is also employed on the linear matrix system. To describe the preconditioner used let

$$A = L + D + U,$$

where D is the diagonal, L is the lower triangular part, and U is the upper triangular part. Then, P_L is used as left preconditioner and P_R as right preconditioner, where

$$P_L = (L + D)^{-1}, \qquad P_R = (D^{-1}U + I)^{-1}.$$

The initial data was chosen to have form

$$u_0(x) = c_1 \sin \pi x + c_2 \sin 2\pi x + c_3 \sin 3\pi x,$$
$$v_0(x) = c_4 \sin \pi x + c_5 \sin 2\pi x,$$

for various constants $c_1, \dots, c_5$.

In Sect. 2.1 we have seen that conditional decay will occur for u_0, v_0 small enough. If we momentarily restrict attention to system A then for the linearised system, i.e. without the uu_x and vv_x terms, we see that the solution

$$u = \hat{u}\, e^{\sigma t} \sin kx, \qquad v = \hat{v}\, e^{\sigma t} \sin kx,$$

leads to the condition

$$\begin{vmatrix} \sigma + k^2/R - \alpha & -\gamma \\ -\delta & \sigma + k^2/R - \beta \end{vmatrix} = 0.$$

From this we find

$$\sigma = \frac{\alpha + \beta - 2k^2/R \pm \sqrt{(2k^2/R - \alpha - \beta)^2 - 4(\alpha\beta - \gamma\delta)}}{2}.$$

If $\gamma\delta > \alpha\beta$ then the square root is real and $\alpha + \beta > 2k^2/R$ ensures linear instability. If $\alpha + \beta < 2k^2/R$ then with $A = 2k^2/R - \alpha - \beta(> 0)$ and $\epsilon = 4(\gamma\delta - \alpha\beta)$ we see that $\sigma = \frac{1}{2}(-A \pm \sqrt{A^2 + \epsilon})$ so with the positive sign $\sigma > 0$ and linear instability again occurs. Thus for linear stability to occur it is necessary that $\alpha\beta > \gamma\delta$ and $A > 0$ for then $\sigma = \frac{1}{2}(-A \pm \sqrt{A^2 - \epsilon})$ and both solutions are negative. Thus, since $x \in (0,1)$, $k_{min} = \pi$ and linear stability can only occur for

$$\alpha\beta > \gamma\delta \qquad \text{and} \qquad \alpha + \beta < \frac{2\pi^2}{R}. \tag{2.5.11}$$

When (2.5.11) does not hold the solution to the linearised version of system A will grow exponentially in time. We ensure that the initial data are large enough that conditional decay will not happen, and we also require condition (2.5.11) to be negated in our numerical calculations.

Before proceeding to the numerical results it is worth noting that an unconditional decay result is easily derived for the full non-linear version of system A by an energy analysis. Multiply (2.5.3)$_1$ by u, (2.5.3)$_2$ by v, and integrate over (0,1) to obtain

$$\frac{d}{dt}\frac{1}{2}\|u\|^2 = -\frac{1}{R}\|u_x\|^2 + \alpha\|u\|^2 + \gamma(u,v),$$

$$\frac{d}{dt}\frac{1}{2}\|v\|^2 = -\frac{1}{R}\|v_x\|^2 + \beta\|v\|^2 + \delta(u,v).$$

Upon addition of these equations with the involvement of a positive coupling parameter λ, we derive

$$\frac{d}{dt}\frac{1}{2}\left(\|u\|^2 + \lambda\|v\|^2\right) = I - \frac{D}{R},$$

where

$$D = \|u_x\|^2 + \lambda\|v_x\|^2,$$
$$I = \alpha\|u\|^2 + (\gamma + \lambda\delta)(u,v) + \lambda\beta\|v\|^2.$$

One then derives

$$\frac{d}{dt}\frac{1}{2}\left(\|u\|^2 + \lambda\|v\|^2\right) \leq -D\left(\frac{1}{R} - \frac{1}{R_E}\right),$$

where

$$\frac{1}{R_E} = \max_H \frac{I}{D},$$

with H being the space of admissible solutions for u and v. Unconditional decay follows if $R < R_E$, cf. Straughan (1992).

To find R_E we first transform $\sqrt{\lambda}v \to v$ and then find the Euler–Lagrange equations for

$$R_E^{-1} = \max_H \frac{\alpha\|u\|^2 + (\gamma + \sqrt{\lambda}\delta)(u,v) + \beta\|v\|^2}{\|u_x\|^2 + \|v_x\|^2}.$$

The Euler–Lagrange equations are

$$R_E\left[2\alpha u + (\gamma + \sqrt{\lambda}\delta)v\right] + 2\frac{d^2 u}{dx^2} = 0,$$

$$R_E\left[2\beta v + (\gamma + \sqrt{\lambda}\delta)u\right] + 2\frac{d^2 v}{dx^2} = 0.$$

Solution of these requires

$$\begin{vmatrix} 2\alpha R_E - 2k^2 & R_E(\gamma + \sqrt{\lambda}\delta) \\ R_E(\gamma + \sqrt{\lambda}\delta) & 2\beta R_E - 2k^2 \end{vmatrix} = 0.$$

Hence

$$R_E^2\left[4\alpha\beta - (\gamma + \sqrt{\lambda}\delta)^2\right] - 4(\alpha + \beta)k^2 R_E + 4k^4 = 0.$$

By consideration of the coefficient of R_E^2 and remembering we want the smallest value of R_E we find

$$RE = \frac{2k^2}{\mathcal{D}} \left[(\alpha + \beta) - \sqrt{(\alpha + \beta)^2 - \mathcal{D}} \right], \qquad (2.5.12)$$

where

$$\mathcal{D} = 4\alpha\beta - (\gamma + \sqrt{\lambda}\delta)^2.$$

To find the optimal value of R_E we find

$$\max_{\lambda} \min_{k^2} R_E$$

with $k = n\pi$, $n = 1, 2, \ldots$. This calculation is not performed here, but we see there is a value of R_E for which $R < R_E$ guarantees unconditional decay for a solution to system A. In the numerical work which follows we are interested in the situation in which the solution develops into a non-zero steady state.

The steady states show that diffusion would appear important only near the right boundary where a steep boundary layer is observed, the solution in the rest of the region behaving like an inviscid one. This behaviour was also found in the single equation cases, Levine *et al.* (1989), Straughan *et al.* (1987), as reported at the beginning of this section. What appears somewhat surprising is that the solution to system A approaches a steady state which consists of two linear functions across the layer before diffusion takes over. This resembles the single equation case, cf. Straughan *et al.* (1987), but there is no *a priori* reason why the interaction should yield the same behaviour in the system. In Fig. 2.1 we display the steady states achieved for various combinations of R, α, β, γ and δ. It is to be noticed that in all six cases the linearity of u and v in x is evident until diffusion becomes important near the boundary $x = 1$.

In Fig. 2.1a we see that the dominant forces are α and β and the solution reflects this with the slopes closely resembling $\alpha + \gamma$ and $\delta + \beta$. Figures 2.1b and 2.1c show the steady states for small α and β with the cross effects being dominant. Again the solution is linear in x until diffusion is important. The higher R value clearly has a steeper boundary layer. It is noteworthy though that the u and v slopes are the same until approximately $x = 0.7$, which indicates the solution is driven by the diffusionless mechanism. In Figs. 2.1d and 2.1e the parameters $\alpha, \beta, \gamma, \delta$ vary by a factor of 10. The solution slopes are essentially the same apart from the factor of 10. Finally, Fig. 2.1f shows the steady state for a situation in which α and δ are largest. The linearity in x is again evident.

In the following discussion of numerical results we concentrate on systems B,C,D,G and H, system A having been dealt with above.

Figure 2.2 contains plots of the steady states for system H with a variety of choices of the parameters $R, \alpha, \beta, \gamma, \delta$. In all pictures it is clear that the solutions u, v are evidently growing exponentially across the layer from $x = 0$ to where diffusion takes over and a boundary layer is present to return the solution to zero at $x = 1$. In Fig. 2.2a the terms α and β are dominant so the solutions resemble those of the single equation cases. Figure 2.2b shows what happens with the "cross forces" being dominant, i.e. the effect of v^2 on the u-equation and the effect of u^2 on the v-equation should dominate. Figures 2.2c and 2.2d demonstrate other values, with Fig. 2.2c employing parameters with similar numerical values.

In Fig. 2.3a α and β are the dominant parameters and this is reflected in the steady states to system B. That for u appearing to increase exponentially while that for v increases linearly. Figure 2.3b shows a similar state of affairs but with a stronger effect on the v-equation. In Fig. 2.3c the effect of α is small and the solutions u and v both appear almost linear over the layer. In Fig. 2.3d the α effect is comparable to the other (β, γ, δ) linear force effects. Interestingly, this case shows both u and v curved even though the v-equation contains only linear froces. The v solution is obviously being influenced by the quadratic u^2 term in the u-equation.

Figures 2.4a-d show the steady states for system C for a selection of parameters. Figure 2.4c shows that a relatively small u^2 and v^2 effect in the u-equation still leads to u having a curved profile. In Fig. 2.4a the v-solution is noticeably curved. Figure 2.5 displays the steady states for system D. These pictures are particularly interesting. In this system the only quadratic force is the γv^2 term acting on the u-equation. The steady states for system G displayed in Figs. 2.6a-d are likewise interesting. This system has "cross effect" quadratic forces via the γv^2 and δu^2 terms, the u force in the u-equation and the v force in the v-equation being linear. Figure 2.6a shows that relatively large "cross" forces yield substantially curved steady states. Figure 2.6d displays that when these quadratic forces are small the profiles are fundamentally linear. In Figs. 2.6b and 2.6c curved profiles for v and u are observed, respectively, despite the large values of the coefficients of the linear forces.

In Figs. 2.1 - 2.6 the pictures are placed so that Fig. Xa is the upper left one, $X = 2.1, \ldots, 2.6$, Fig. Xb is the upper right one, Fig. Xc is the one on the left in the second row, and so on.

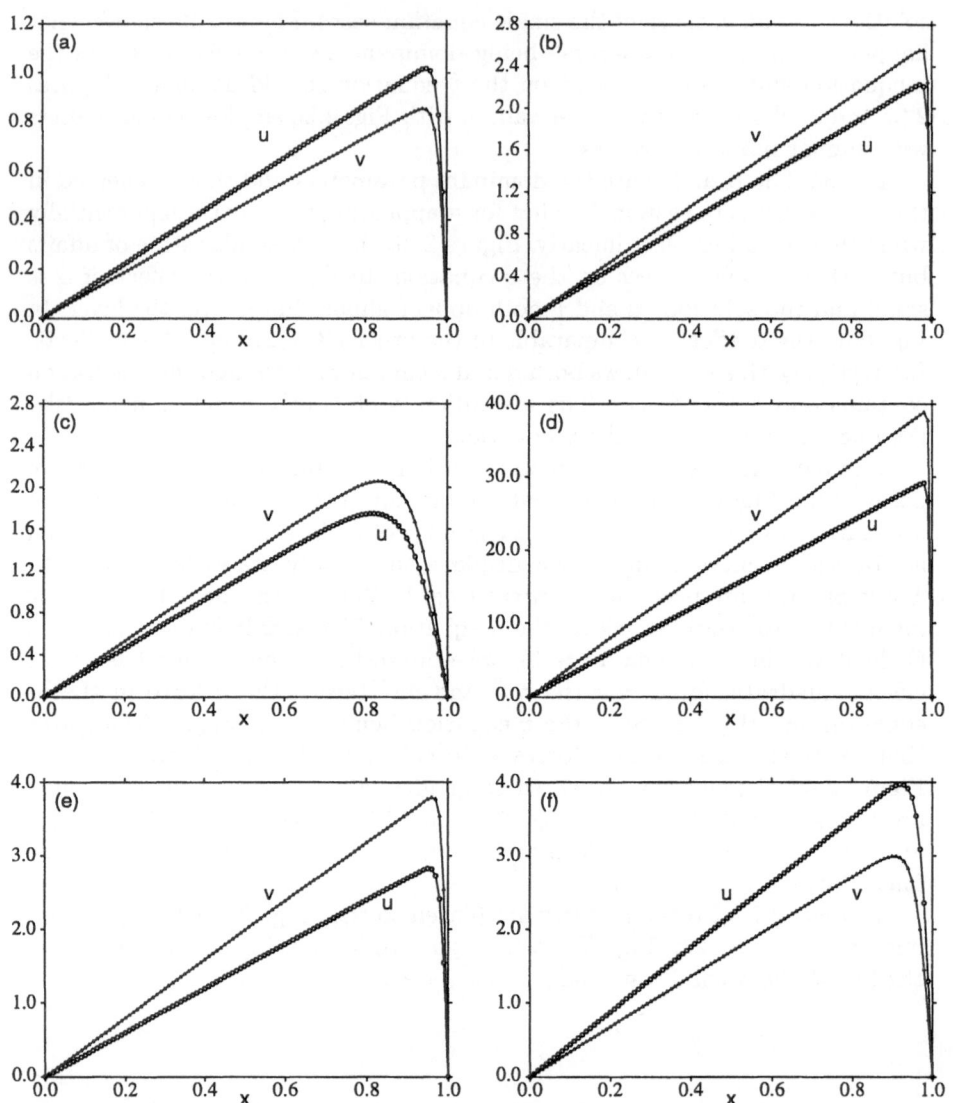

Fig. 2.1 Steady solutions to system A. (a) $R = 100, \alpha = 1, \gamma = 0.1, \delta = 0.1, \beta = 0.8$; (b) $R = 100, \alpha = 0.01, \gamma = 2, \delta = 3, \beta = 0.02$; (c) $R = 10, \alpha = 0.01, \gamma = 2, \delta = 3, \beta = 0.02$; (d) $R = 10, \alpha = 10, \gamma = 15, \delta = 50, \beta = 2$; (e) $R = 40, \alpha = 1, \gamma = 1.5, \delta = 5, \beta = 0.2$; (f) $R = 15, \alpha = 4, \gamma = 0.5, \delta = 2, \beta = 0.8$.

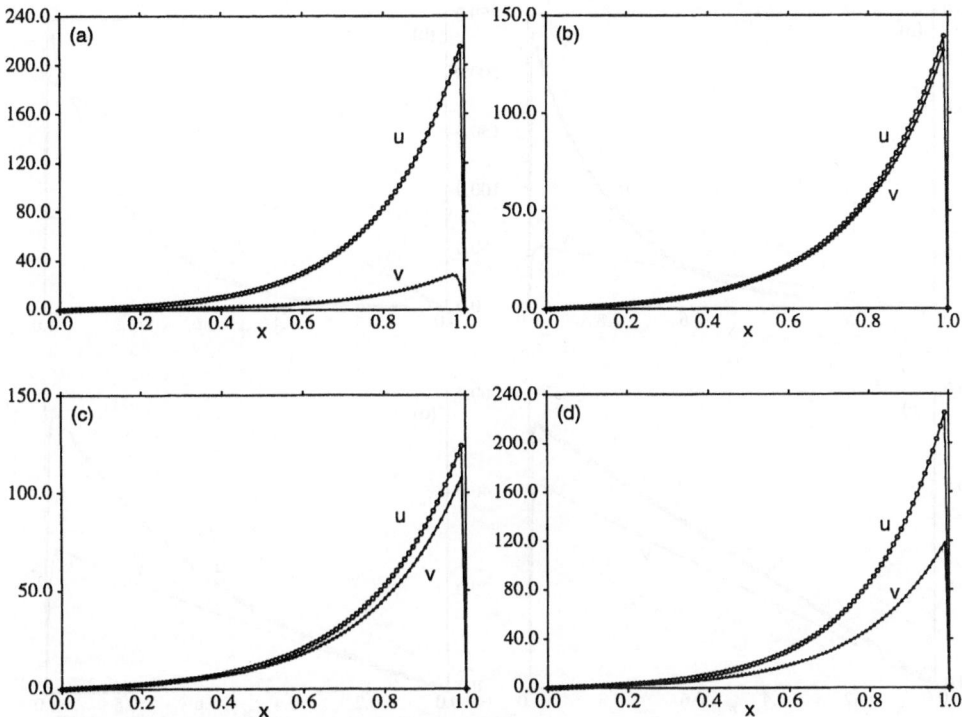

Fig. 2.2 Steady solutions to system H. (a) $R = 5, \alpha = 5, \gamma = 0.1, \delta = 0.1, \beta = 0.1$; (b) $R = 5, \alpha = 0.1, \gamma = 5, \delta = 4, \beta = 0.2$; (c) $R = 5, \alpha = 3, \gamma = 2, \delta = 1.5, \beta = 2.5$; (d) $R = 5, \alpha = 5, \gamma = 0.1, \delta = 0.2, \beta = 4$.

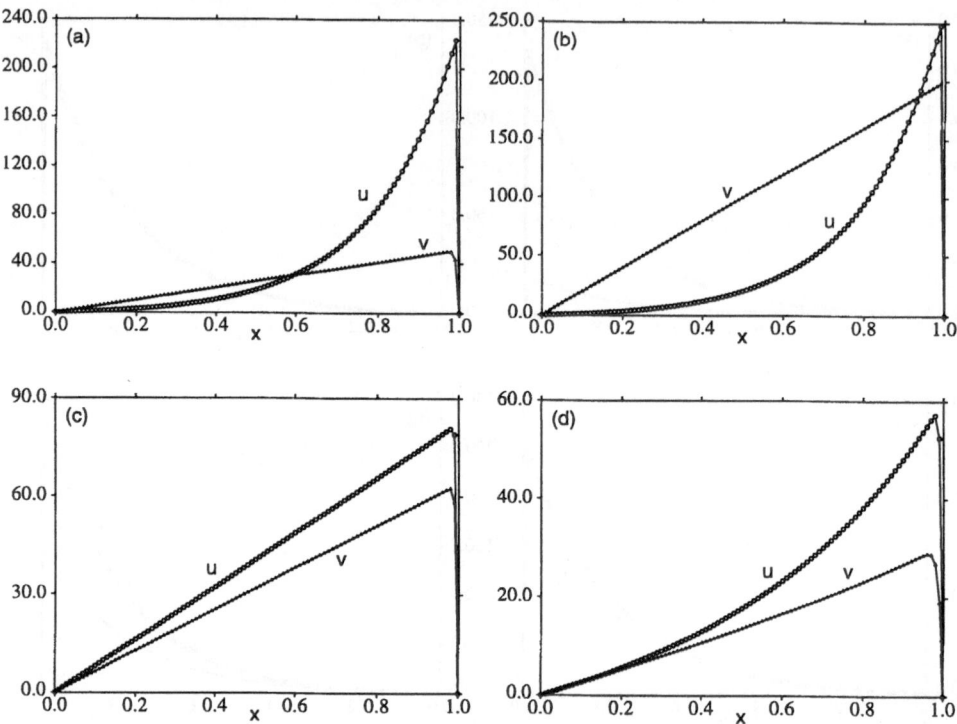

Fig. 2.3 Steady solutions to system B. (a) $R = 5, \alpha = 5, \gamma = 0.1, \delta = 1, \beta = 50$; (b) $R = 5, \alpha = 5, \gamma = 0.1, \delta = 1, \beta = 200$; (c) $R = 5, \alpha = 0.1, \gamma = 100, \delta = 50, \beta = 0.3$; (d) $R = 5, \alpha = 2, \gamma = 20, \delta = 15, \beta = 10$.

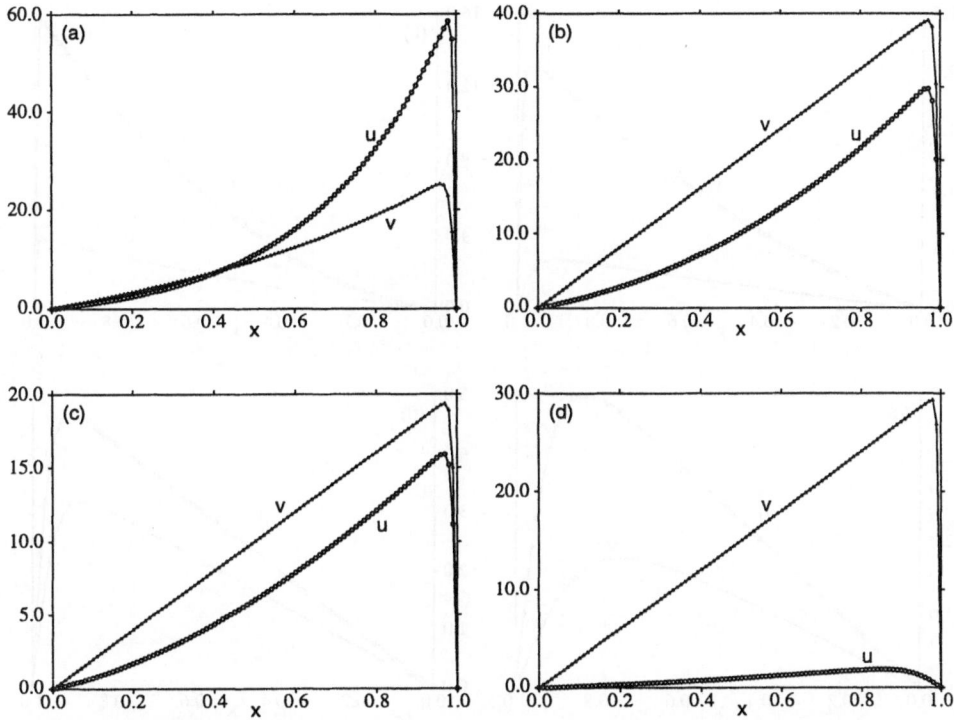

Fig. 2.4 Steady solutions to system C. (a) $R = 5, \alpha = 3, \gamma = 1, \delta = 20, \beta = 2$; (b) $R = 5, \alpha = 1, \gamma = 0.5, \delta = 1, \beta = 40$; (c) $R = 10, \alpha = 0.1, \gamma = 1, \delta = 0.1, \beta = 20$; (d) $R = 10, \alpha = 0.01, \gamma = 0.01, \delta = 1, \beta = 30$.

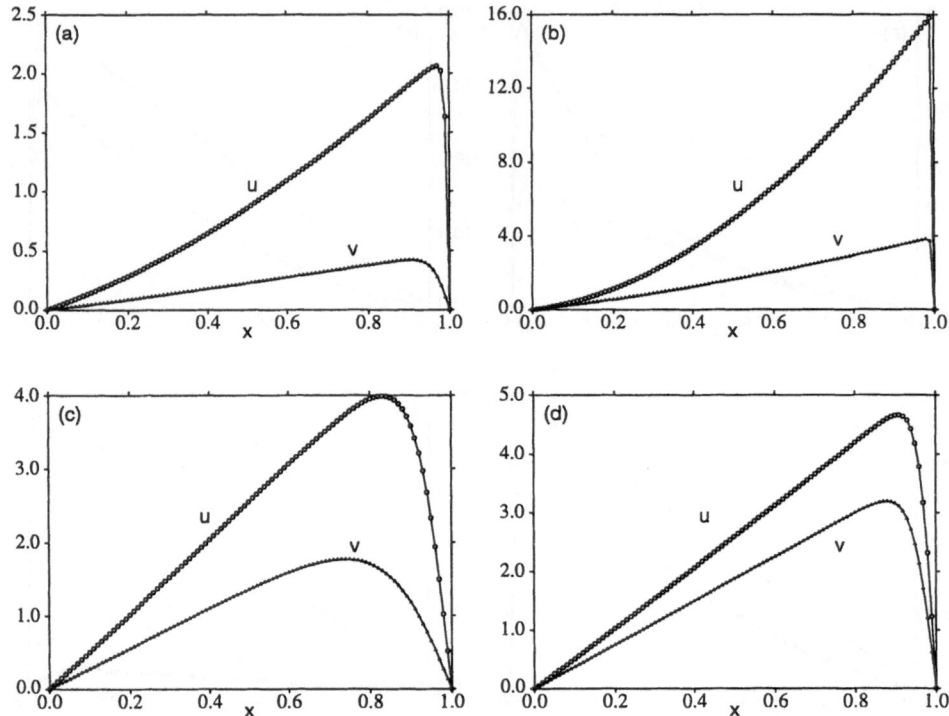

Fig. 2.5 Steady solutions to system D. (a) $R = 100, \alpha = 1, \gamma = 10, \delta = 0.1, \beta = 0.1$; (b) $R = 100, \alpha = 0.01, \gamma = 15, \delta = 1, \beta = 1$; (c) $R = 5, \alpha = 5, \gamma = 0.1, \delta = 1, \beta = 1$; (d) $R = 10, \alpha = 5, \gamma = 0.1, \delta = 2, \beta = 1$.

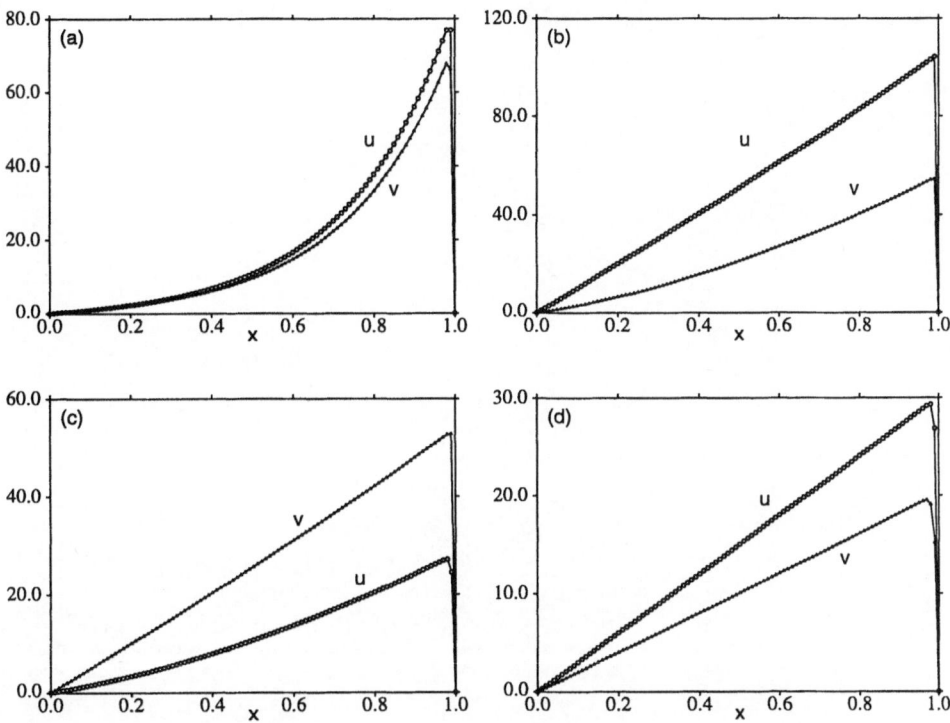

Fig. 2.6 Steady solutions to system G. (a) $R = 5, \alpha = 1, \gamma = 5, \delta = 3, \beta = 2$; (b) $R = 10, \alpha = 100, \gamma = 0.5, \delta = 0.3, \beta = 20$; (c) $R = 10, \alpha = 10, \gamma = 0.3, \delta = 0.7, \beta = 50$; (d) $R = 10, \alpha = 30, \gamma = 0.01, \delta = 0.01, \beta = 20$.

3. Singularities for Classical Fluid Equations

3.1 Breakdown for First-Order Systems

Levine & Protter (1986) present an interesting analysis for the system

$$\frac{\partial u_i}{\partial t} + f_m(x,t,u)\frac{\partial u_i}{\partial x_m} + g_i(x,t,u) = 0, \qquad (3.1.1)$$

where $i = 1, \ldots, n$, and the repeated index denotes summation from 1 to n. This equation is studied for $x \in \mathbf{R}^n$ with u_i satisfying the initial data

$$u_i(x,0) = u_i^0(x), \qquad i = 1, \ldots, n; \quad x \in \mathbf{R}^n. \qquad (3.1.2)$$

The work of Levine & Protter (1986) is very interesting in that it relates the behaviour of a solution to (3.1.1) to a solution of a second-order linear system. To do this they differentiate (3.1.1) with respect to x_j and with repeated indices denoting summation from 1 to n this gives

$$\frac{\partial^2 u_i}{\partial x_j \partial t} + f_r\frac{\partial^2 u_i}{\partial x_r \partial x_j} + \frac{\partial f_r}{\partial x_j}\frac{\partial u_i}{\partial x_r} + \frac{\partial f_r}{\partial u_m}\frac{\partial u_m}{\partial x_j}\frac{\partial u_i}{\partial x_r}$$

$$+ \frac{\partial g_i}{\partial x_j} + \frac{\partial g_i}{\partial u_m}\frac{\partial u_m}{\partial x_j} = 0. \qquad (3.1.3)$$

Then with the notation

$$J = \frac{\partial u_i}{\partial x_j}, \quad \nabla_x f = \frac{\partial f_i}{\partial x_j}, \quad \nabla_u f = \frac{\partial f_i}{\partial u_j},$$

and denoting the total derivative D/Dt by

$$\frac{D}{Dt} = \frac{\partial}{\partial t} + f_m(x,t,u)\frac{\partial}{\partial x_m},$$

they write (3.1.1) and (3.1.3), respectively, as

$$\frac{Du}{Dt} + g(x,t,u) = 0, \qquad (3.1.4)$$

$$\frac{DJ}{Dt} + J\nabla_x f + J\nabla_u fJ + \nabla_x g + \nabla_u gJ = 0, \qquad (3.1.5)$$

where the coefficient matrices are understood to be evaluated along the characteristics associated with (3.1.1). They transform to a variable Y by putting

$$J = (\nabla_u f)^{-1} Y'(t) Y^{-1}(t), \tag{3.1.6}$$

and from (6.4.5) find Y satisfies

$$[(\nabla_u f)^{-1} Y']' + \nabla_u g (\nabla_u f)^{-1} Y'$$
$$+ (\nabla_x g) Y + (\nabla_u f)^{-1} Y' Y^{-1} (\nabla_x f) Y = 0. \tag{3.1.7}$$

This is a second-order matrix differential equation for Y and Levine & Protter (1986) investigate the relationship between a solution to (3.1.1) and a solution to this equation. They first suppose

$$f_i = f_i(t, u), \qquad g_i = g_i(t), \tag{3.1.8}$$

and are then able to show:

Theorem. (Levine & Protter (1986)).
Suppose f and g satisfy (3.1.8) along a characteristic. If $J(0)$ has a negative eigenvalue and if the eigenvalues of the matrix

$$\int_0^t \nabla_u f \left(s, u(0) + \int_0^s g(p)\,dp \right) ds$$

grow without bound as $t \to \infty$, then there is a $T > 0$ such that $J(t)$ becomes unbounded as $t \to T^-$ in the norm

$$\|J(t)\| = \max_{i,j=1,\dots,n} |u(x(t), t)|.$$

They also make interesting deductions with the case

$$f = f(t, u), \qquad g = g(t, x). \tag{3.1.9}$$

For this case (3.1.5) has form

$$\frac{DJ}{Dt} = -J \nabla_u f J - \nabla_x g, \tag{3.1.10}$$

and so by putting

$$P = \nabla_u f, \qquad Q = \nabla_x g,$$

this becomes, provided J^{-1} exists,

$$\frac{D}{Dt}(J^{-1}) = P + J^{-1} Q J^{-1}. \tag{3.1.11}$$

Levine & Protter (1986) are then led to study the system of first-order equations

$$\frac{dx}{dt} = -Q(t)y(t),$$
$$\frac{dy}{dt} = P(t)x(t). \tag{3.1.12}$$

The relationship between a solution to (3.1.11) and a solution to (3.1.12) really is quite interesting. To understand this one requires some definitions.

The points $a, b \in \mathbf{R}^+ \cup \{0\}$ are called *conjugate points* provided there is a solution (x, y) to (3.1.12) such that $y \neq 0$ for $t \in (a, b)$ and $y(a) = y(b) = 0$. System (3.1.12) is *oscillatory* on $\mathbf{R}^+ \cup \{0\}$ if for every $a \in \mathbf{R}^+, \exists b > a$ such that a and b are conjugate points. Levine & Protter (1986) relate the behavior of a solution to (3.1.10) to the oscillatory characteristics of a system by employing a theorem due to Rasmussen (1979):

Theorem. (Rasmussen (1979)).
Let $P(t), Q(t)$ be Hermitian positive semidefinite matrices on $\mathbf{R}^+ \cup \{0\}$. Then the system (3.1.12) is oscillatory if and only if the adjoint system

$$\frac{du}{dt} = -P(t)v(t),$$
$$\frac{dv}{dt} = Q(t)u(t), \tag{3.1.13}$$

is oscillatory on $\mathbf{R}^+ \cup \{0\}$. Equivalently, solutions to (3.1.13) are oscillatory if and only if every Hermitian, absolutely continuous non-singular solution to (3.1.11) becomes singular at some finite time $T > 0$. Furthermore, every Hermitian, absolutely continuous non-singular solution to (3.1.11) becomes singular in finite time if and only if every Hermitian, absolutely continuous solution to

$$\frac{DJ}{Dt} = -Q - JPJ$$

becomes singular for some positive time T'.

The precise theorem of Levine & Protter (1986) is:

Theorem. (Levine & Protter (1986)).
Let $\nabla_x f = \nabla_u g = 0$ along a characteristic, and suppose that $\nabla_u f$ and $\nabla_x g$ are Hermitian and positive definite along this characteristic. Suppose that $J(0)$, the initial values of $J(t)$, is Hermitian and that solutions to (3.1.13) are oscillatory. Then there is a $T > 0$ such that

$$\lim_{t \to T} \|J(t)\| = \infty. \tag{3.1.14}$$

Conversely, suppose that for every set of Hermitian initial values $J(0)$, the solution along the characteristic becomes unbounded at a finite time as given by (3.1.14). Then every solution to (3.1.13) is oscillatory.

Levine & Protter (1986) also extend this to give conditions under which solutions to (3.1.1) must develop a singularity in the first derivatives in a finite time. They relate this behavior to systems of equations in gas dynamics.

3.2 Blow-Up of Solutions to the Euler Equations

Whether singularities form in finite time for the equations of hydrodynamics with a linear constitutive law is a topic engaging many mathematicians and physicists. We here describe some of the recent findings in this field.

The question of breakdown of the solution to Euler's equations, even for smooth initial data, is an extremely important one, and many writers are studying this, see e.g. Beale *et al.* (1984), Constantin *et al.* (1994), Ferrari (1993), Majda (1986), and the references therein. General results are available, such as that of Beale *et al.* (1984) who study Euler's equations on $\mathbf{R}^3$, i.e.

$$\frac{\partial u_i}{\partial t} + u_j \frac{\partial u_i}{\partial x_j} = -\frac{\partial p}{\partial x_i},$$
$$\frac{\partial u_i}{\partial x_i} = 0. \tag{3.2.1}$$

They deal specifically with the vorticity

$$\omega_i = (\nabla \times u)_i$$

and show that if u_i is a solution to Euler's equations, and there is a time (a first such time) T such that u_i cannot be continued in the class

$$C\left([0,T]; H^s\right) \cap C^1\left([0,T]; H^{s-1}\right),$$

then the vorticity must be unbounded in the sense that

$$\int_0^T |\omega(t)|_{L^\infty} dt = \infty,$$

and

$$\lim_{t \to T^-} \sup |\omega(t)|_{L^\infty} = \infty.$$

The work of Beale *et al.* (1984) on singularity formation for a solution to the Euler equations is also discussed by Doering & Gibbon (1995), pp. 153–157. The subject of existence/non-existence is further discussed in detail in Majda (1986); he points out there is numerical evidence for finite time breakdown, but to our knowledge the general question remains as it was at the time of writing Majda (1986) and Constantin *et al.* (1994), and has not yet been resolved. Indeed, Constantin *et al.* (1994) write "*Whether the three-dimensional Euler equations develop finite time singularities from smooth initial data is an important and highly controversial open problem.*"

Ferrari (1993) studies the equivalent problem to that of Beale *et al.* (1984) but when the spatial domain Ω is a bounded domain in $\mathbf{R}^3$. The boundary conditions chosen are those for which $u_i n_i$ is zero on the boundary of Ω. Ferrari (1993) shows that if $[0,\hat{T})$ is the maximal interval of existence of a solution u_i to (3.2.1) with $u_i n_i$ zero on the boundary, and with $u_i \in C([0,T]; H^s(\Omega))$, with $s \geq 3$ and $T < \hat{T}$, then

$$\int_0^{\hat{T}} |\omega(\cdot, t)|_{L^\infty(\Omega)} dt = \infty,$$

and

$$\sup_{t \in [0,\hat{T})} |\omega(\cdot, t)|_{L^\infty(\Omega)} = \infty.$$

The proof of Ferrari's result is necessarily very different from that when $\Omega \equiv \mathbf{R}^3$ and is of much interest in its own right, employing elliptic estimates in a neat manner.

An appealing procedure to establish blow-up in fluid dynamics has been developed by Stuart (1988, 1991), see also Stuart & Tabor (1990). This treats the Euler equations specifically from a Lagrangian coordinate viewpoint. By considering special flows explicit information may be derived, e.g. Stuart (1988) considers the flow near a plane of symmetry $z = 0$ and takes the velocity components $\mathbf{u} = (u, v, w)$ and pressure p as

$$\begin{aligned} u(\mathbf{x}, t) &= axu(y, t), \\ v(\mathbf{x}, t) &= v(y, t), \\ w(\mathbf{x}, t) &= bzw(y, t), \\ p(\mathbf{x}, t) &= -\frac{1}{2}a^2 x^2 - \frac{1}{2}b^2 z^2 p_2 + p(y, t), \end{aligned} \qquad (3.2.2)$$

for a, b constants. These expressions in the Euler equations (3.2.1) yield the system of partial differential equations

$$\begin{aligned} \frac{\partial u}{\partial t} + v\frac{\partial u}{\partial y} + a(u^2 - 1) &= 0, \\ \frac{\partial v}{\partial t} + v\frac{\partial v}{\partial y} + \frac{\partial p}{\partial y} &= 0, \\ \frac{\partial w}{\partial t} + v\frac{\partial w}{\partial y} + b(w^2 - p_2) &= 0, \\ au + \frac{\partial v}{\partial y} + bw &= 0, \end{aligned} \qquad (3.2.3)$$

with the initial and boundary conditions:

$$\begin{aligned} u &= u_0(y), \quad w = w_0(y), \qquad \text{at} \quad t = 0, \\ v &= 0, \qquad \text{when} \quad y = 0, \\ u &\to 1, \quad w \to 0, \qquad \text{as} \quad y \to \infty. \end{aligned}$$

He shows, in fact, that

$$u = \frac{\tanh at + u_0(\phi)}{1 + u_0(\phi)\tanh at},$$

$$w = \frac{w_0(\phi)}{1 + btw_0(\phi)},$$

(3.2.4)

$$v = \frac{\partial Y}{\partial t}(t, \phi),$$

where ϕ is a characteristic variable and

$$Y = \int_0^\phi \frac{d\xi}{[1 + btw_0(\xi)][\cosh at + u_0(\xi)\sinh at]}.$$

Depending on the initial profiles, Stuart (1986) shows that the solution can blow up in finite time, v and w behaving like

$$v = \frac{1}{(T - t)^{3/2}}V(t, \phi),$$

$$w = \frac{1}{(T - t)}W(t, \phi),$$

with V, W finite as $t \to T$.

Other papers which consider special classes of solution to Euler's equations are those of Childress *et al.* (1989), and Jacqmin (1991); a review of related material is also given by Drazin (1991). Childress *et al.* (1989) study a class of two-dimensional solutions to Euler's equations of form

$$(u, v) = \left(f(x, t), -y\frac{\partial f}{\partial x}(x, t)\right),$$

which when put into the vorticity equation which is derived by taking the curl of (3.2.1), namely

$$\frac{\partial \omega}{\partial t} + u\frac{\partial \omega}{\partial x} + v\frac{\partial \omega}{\partial y} = 0,$$

(3.2.5)

yields the equation

$$\frac{\partial^3 f}{\partial t \partial x^2} + f\frac{\partial^3 f}{\partial x^3} - \frac{\partial f}{\partial x}\frac{\partial^2 f}{\partial x^2} = 0.$$

(3.2.6)

This equation may be integrated to yield

$$\frac{\partial^2 f}{\partial t \partial x} + f\frac{\partial^2 f}{\partial x^2} - \left(\frac{\partial f}{\partial x}\right)^2 = h(t),$$

(3.2.7)

where

$$h(t) = -\frac{2}{L}\int_0^L \left(\frac{\partial f}{\partial x}\right)^2 dx.$$

The boundary conditions are

$$f(0,t) = f(L,t) = 0,$$
$$\frac{\partial f}{\partial x}(0,t) = \frac{\partial f}{\partial x}(L,t) = 0.$$

Childress *et al.* (1989) give three proofs of blow-up of a solution to (3.2.7), two of these relying on a transformation to Lagrangian variables. The third seeks a solution like

$$f(x,t) = \frac{F(x)}{T-t}$$

and reduces to an equation for F, namely

$$F' + FF'' - (F')^2 = (T-t)^2 h(t),$$

with $F(0) = F(L) = 0$. The equation for F is studied in detail. The paper by Childress *et al.* (1989) also investigates numerically a viscous analogue of (3.2.7), but we understand from a personal communication from Professor Ierley that the blow-up question in the viscous case is still open.

Jacquim (1991) used methods not dissimilar to those of Stuart (1988) and Childress *et al.* (1989), in an investigation of flow of a variable density fluid in a porous channel of finite height. The equations here are

$$\frac{\partial u}{\partial x} + \frac{\partial w}{\partial z} = 0,$$
$$u = -\frac{\partial p}{\partial x},$$
$$w = -\left(\frac{\partial p}{\partial z} + \rho\right), \tag{3.2.8}$$
$$\frac{\partial \rho}{\partial t} + u\frac{\partial \rho}{\partial x} + w\frac{\partial \rho}{\partial z} = R\left(\frac{\partial^2 \rho}{\partial x^2} + \frac{\partial^2 \rho}{\partial z^2}\right),$$

for R constant, with ρ being the density. The boundary conditions are

$$w = \frac{\partial \rho}{\partial z} = 0, \qquad \text{at} \quad z = \pm 1,$$
$$u = \frac{\partial \rho}{\partial x} = 0, \qquad \text{at} \quad x = 0.$$

Jacquim (1991) chooses

$$\rho = \rho_0 + \rho_2 x^2, \qquad p = p_0 + p_2 x^2,$$
$$u = u_1 x \qquad \text{and} \qquad w = w_0,$$

the subscripted variables being functions of z and t. The resulting system of equations is then reduced to an equation for w_0, namely:

$$\frac{\partial^2 w_0}{\partial t \partial z} + w_0 \frac{\partial^2 w_0}{\partial z^2} - \frac{3}{2}\left(\frac{\partial w_0}{\partial z}\right)^2 = h(t), \tag{3.2.9}$$

with

$$h(t) = -\frac{5}{4} \int_{-1}^{1} \left(\frac{\partial w_0}{\partial z}\right)^2 dz.$$

An argument involving convexity, monotonicity and a comparison function is used to establish blow up of $\partial w_0/\partial z$ with the time of blowing up, T, being estimated by

$$T < \frac{1}{0.6107 \left(\partial w_0/\partial z\right)\Big|_{z=-1,\, t=0}}.$$

Another approach to blow-up of solutions to Euler's equations is described by Novikov (1990). This requires the velocity field to be linear in the spatial variables so that $v_{ij} = \partial v_i/\partial x_j$ is a function only of t. Novikov's (1990) analysis differentiates (3.2.1) with respect to x_j to obtain

$$\frac{\partial v_{ij}}{\partial t} + v_k \frac{\partial v_{ij}}{\partial x_k} + v_{ik}v_{kj} = -p_{ij}, \qquad (3.2.10)$$

where $p_{ij} = \partial^2 p/\partial x_i \partial x_j$, and then from incompressibility and symmetry arguments it is deduced that

$$p_{ij} = -a_{ijklmn}(t)v_{kl}v_{mn}, \qquad (3.2.11)$$

with

$$a_{iiklmn} = \delta_{kn}\delta_{lm}; \quad a_{ijklmn} = a_{jiklmn} = a_{ijmnkl}.$$

Equations (3.2.10) and (3.2.11) lead to a non-linear equation for v_{ij}. The resulting equation is broken down into the symmetric and skew symmetric parts for v_{ij} and then explicit solutions are obtained from which blow-up in finite time may be exhibited.

3.2.1 Vortex Sheet Breakdown and Rayleigh–Taylor Instability

There are many instances of instabilities in a fluid interface, e.g. instabilities in a fluid surface, an instability at the interface when a heavier fluid overlies a lighter fluid (the Rayleigh–Taylor instability), the instability generated at the interface by shearing of one fluid relative to another (Kelvin–Helmholtz instability), and several have wide practical application. Moore's (1979) analysis predicts that a singularity may form in a vortex sheet in a finite time and there are many subsequent papers dealing with precisely this topic. The paper by Baker et al. (1993) is a very interesting one and gives many references to other articles discussing the subject, together with a brief historical development of the theory.

The evolution problem for a flat vortex sheet is known to lead to finite time blow-up of the curvature, cf. Moore (1979). The question of the behaviour of the vortex sheet after the singularity has developed is still open.

The paper by Baker et al. (1993) establishes similar formation of a singularity in the Rayleigh–Taylor problem. These writers use an analytic approach via asymptotics in addition to numerical simulation, and are so able to judge the validity of the asymptotic theory.

Baker *et al.* (1993) study the problem of a heavy fluid overlying a lighter one with gravity being the destabilizing ingredient. The motion is taken to be two-dimensional and equations are derived for the interface position, z, and the strength, γ, of the vortex sheet along the interface. The strength is connected to the quantity

$$\tilde{\gamma} = (\mathbf{u}_1 - \mathbf{u}_2) \cdot \hat{\mathbf{s}},$$

where $\hat{\mathbf{s}}$ is the unit tangent vector to the interface with $\mathbf{u}_1, \mathbf{u}_2$ being the velocities in the lower and upper fluids, respectively, evaluated at the interface. A set of non-linear partial integro-differential equations is derived, cf. Baker *et al.* (1982), to describe the evolutionary development of z and γ. The mathematical analysis is technical but beautiful and enables Baker *et al.* (1993) to predict where and when a singularity will form in finite time. The results predicted by the exact solutions to the set of approximate (asymptotic) equations are in close agreement with numerical simulations on the full equations unless the Atwood number,

$$A = \frac{\rho_2 - \rho_1}{\rho_2 + \rho_1}$$

is close to unity.

This is a problem of practical importance which requires the use of many techniques of mathematical analysis and is a rapidly developing area of singularity formation as may be judged by the many recent papers and preprints quoted in Baker *et al.* (1993).

In work not unrelated to the material of this subsection, Tanveer (1996) presents a detailed asymptotic analysis of the Hele–Shaw problem of flow of a less viscous fluid through a more viscous one. In certain cases the problem is ill posed and a finite time singularity is observed on the boundary of the "finger" of less viscous fluid. Tanveer's (1996) paper contains many references to analyses of Hele–Shaw flow.

3.2.2 A Mathematical Theory for Sonoluminescence

The paper of Wu & Roberts (1994) contains an extremely interesting analysis of an important physical problem. They consider the experimental situation where a spherical container of water has a bubble of air trapped at its centre. The surface of the sphere (container) emits sound waves due to transducers placed there. These sound waves give rise to expansion and contraction at the surface of the bubble of trapped air. The oscillations of the surface may in turn generate shock waves in the air in the bubble and under the right conditions these shocks can ionize the air in a neighbourhood of the centre. The ionized air then may emit light due to a phenomenon Wu & Roberts

(1994) called sonluminescence, although they attribute the emission to an effect called bremsstrahlung.

To capture the above effect mathematically Wu & Roberts (1994) begin with the Rayleigh–Lamb equation for the radius of the spherical air bubble, namely

$$R\frac{d^2R}{dt^2} + \frac{3}{2}\left(\frac{dR}{dt}\right)^2 = \frac{1}{\rho_w}[p(R,t) - P_\infty] + \frac{R}{\rho_w c_w}\frac{d}{dt}[p(R,t) - P_\infty]$$
$$- \frac{4\nu}{R}\frac{dR}{dt}. \tag{3.2.12}$$

In this equation ρ_w, c_w and ν are the density of water, speed of sound in the water and kinematic viscosity of the water, respectively. The quantity $p(R,t)$ is the pressure of air at the bubble surface. More generally, $p(r,t)$ is the pressure of water in the bubble. Also, $P_\infty(t)$ is the pressure of the water "at infinity" (in the experiment this is the pressure at the wall of the spherical container). The conditions Wu & Roberts (1994) adopt for P_∞ to model the sonoluminescence experiment are that

$$P_\infty = P_0 + P_a(t), \qquad P_a(t) = -P_a' \sin\omega_a t,$$

P_0, P_a' and ω_a being suitable constants.

Wu & Roberts (1994) contains an interesting historical review of models for the pressure in (3.2.12), since previous workers essentially assumed some model for $p(r,t)$. Wu & Roberts (1994), however, couple (3.2.12) directly to Euler's equations for non-isothermal, non-constant density flow in the bubble itself and thereby determine $p(R,t)$. Thus, they solve the spherically symmetric equations

$$\frac{\partial\rho}{\partial t} + \frac{1}{r^2}\frac{\partial}{\partial r}(\rho v r^2) = 0,$$
$$\frac{\partial}{\partial t}(\rho v) + \frac{1}{r^2}\frac{\partial}{\partial r}(\rho v^2 r^2) + \frac{\partial p}{\partial r} = 0, \tag{3.2.13}$$
$$\frac{\partial E}{\partial t} + \frac{1}{r^2}\frac{\partial}{\partial r}[(E+p)vr^2] = 0.$$

Here ρ, v and p are the density, radial velocity and pressure in the bubble with E being the total density defined by

$$E = \frac{1}{2}\rho v^2 + \rho e. \tag{3.2.14}$$

The quantity e is given by

$$e = c_v T = \left(\frac{V-b}{\gamma-1}\right)p, \tag{3.2.15}$$

in which c_v is the specific heat at constant volume, V is the volume of gas, T is the temperature and p is governed by the van der Waals equation of state

$$p = \frac{\mathcal{R}T}{V - b},$$
(3.2.16)

$\mathcal{R}$ being the gas constant. To solve (3.2.12) - (3.2.16) when shock waves are present Wu & Roberts (1994) introduce an appropriate form of Rankine–Hugoniot conditions.

Although Wu & Roberts (1994) contains a full numerical solution of (3.2.12)–(3.2.16) they compare their results with a similarity solution which had been previously found for an ideal gas. In addition to solving for the fluid thermo - mechanical quantities involved in (3.2.12)–(3.2.16) they also calculate the luminosity (light radiation that the bubble emits). They calculate the full emission via a luminosity function $L(t)$, and a reduced luminosity $L_w(t)$ which allows only for wavelengths of over 200nm, the other radiation being absorbed by the water. The radiation emitted by the ionized air is assumed bremsstrahlung and then to calculate $L(t)$ and $L_w(t)$ Wu & Roberts (1994) integrated the bremsstrahlung power emission per unit volume, P_{Br}, throughout the volume of the bubble (a suitably restrained version for the reduced emission case $L_w(t)$). The quantity P_{Br} is given by

$$P_{Br} = 1.57 \times 10^{-40} q^2 N^2 T^{1/2} \ \mathrm{Wm}^{-3}$$

where q is the degree of ionization which they take from Saha's formula

$$\frac{q^2}{1 - q} = 2.4 \times 10^{21} \frac{T^{3/2}}{N} \, \mathrm{e}^{-\chi/kT} \ .$$

In these equations k is Boltzmann's constant, N is the number density of atoms, and χ is the ionization potential.

Wu & Roberts (1994) solve (3.2.12)–(3.2.16) numerically with a Lax–Friedrichs shock capturing scheme. They employ a moving spatial grid with various grid resolutions to verify convergence of their solutions.

The results discussed in detail by Wu & Roberts (1994) are for four representative cases of the parameters P'_a, ω_a and initial bubble radius R_0. The findings of Wu & Roberts (1994) are very interesting although too detailed to describe at length here. However, they observe conditions under which the temperature and density of the gas at the centre of the bubble exhibits a very sharp spike reminiscent of focussing and finite time blow-up. (Indeed the theory according the to the similarity solution employed by Wu & Roberts (1994) allows for such finite time blow-up behaviour). Wu & Roberts (1994) do exhibit parameters for which shock waves are generated and for which light flashes can be emitted. They do, therefore, present a highly interesting model and analysis for the phenomenon of sonoluminescence. However, they discuss ways in which their model is deficient and how it may possibly be improved.

3.3 Blow-Up of Solutions
to the Navier–Stokes Equations

We remarked at the beginning of the last section that the question of break-down of the solution to Euler's equations, even for smooth initial data, is an extremely important one, and many writers are studying this. The associated question of breakdown to solutions to the Navier–Stokes equations is probably more important.

Very interesting results on finite time blow up for solutions in boundary layers which are governed by the Navier–Stokes equations have recently been presented by Hall *et al.* (1992), Stewart & Smith (1992) and Smith & Bowles (1992). These papers are concerned with important questions of transistion of boundary layer flows into turbulent boundary layers. For example, Stewart & Smith (1992) start with the unsteady three-dimensional interactive boundary layer equations

$$\frac{\partial u}{\partial x} + \frac{\partial v}{\partial y} + \frac{\partial w}{\partial z} = 0,$$

$$\frac{\partial u}{\partial t} + u\frac{\partial u}{\partial x} + v\frac{\partial u}{\partial y} + w\frac{\partial u}{\partial z} = -\frac{\partial p}{\partial x} + \frac{\partial^2 u}{\partial y^2}, \qquad (3.3.1)$$

$$\frac{\partial w}{\partial t} + u\frac{\partial w}{\partial x} + v\frac{\partial w}{\partial y} + w\frac{\partial w}{\partial z} = -\frac{\partial p}{\partial z} + \frac{\partial^2 w}{\partial y^2},$$

subject to the boundary conditions

$$\begin{aligned} u = v = w = 0 &\quad \text{at} \quad y = 0, \\ u \sim y + A(x,z,t), &\quad w \to 0 \quad \text{as} \quad y \to \infty, \end{aligned} \qquad (3.3.2)$$

which represent the no slip condition at a fixed surface and interaction with the flow outside the boundary layer; there is another boundary condition which relates to the pressure. By considering a vortex wave interaction they derive a system of equations which mathematically may be written

$$\frac{1}{2}\frac{\partial \rho}{\partial T} + \frac{\partial}{\partial Z}\left(\rho\frac{\partial \theta}{\partial Z}\right) = \rho,$$

$$\rho^2\frac{\partial \theta}{\partial T} - \frac{1}{2}\rho\frac{\partial^2 \rho}{\partial Z^2} + \frac{1}{4}\left(\frac{\partial \rho}{\partial Z}\right)^2 + \rho^2\left(\frac{\partial \theta}{\partial Z}\right)^2 = -\rho^2 Q, \qquad (3.3.3)$$

$$\frac{\partial^2 Q}{\partial T^2} = \frac{\partial^2 \rho}{\partial Z^2}.$$

To study blow-up they employ a sort of renormalization argument and take

$$\rho \sim \frac{1}{(T_s - T)^2} \, \tilde{\rho}(\eta),$$

$$\theta \sim \frac{\tilde{\lambda}^2 K}{(T_s - T)^3} + \tilde{\theta}(\eta),$$

$$Q \sim \frac{\tilde{\lambda}^2}{(T_s - T)^4} \, \tilde{Q}(\eta),$$

with

$$\eta = \frac{\tilde{\lambda}(Z - Z_s)}{(T_s - T)^2},$$

T_s being the blow-up time and Z_s being the spatial blow-up point. This analysis, in fact, compares the analytical results obtained with experimental work.

A lucid study of the behaviour of a solution to the Proudman–Johnson equation (defined below in (3.3.6)) for flow in a channel may be found in Cox (1991). Cox describes two-dimensional flow of an incompressible linear viscous fluid in a channel, $x \in \mathbf{R}, y \in (-h, h)$, and refers to this as the Berman problem. He introduces a stream function $\psi(x, y, t)$ in the usual way, namely

$$(u, v) = \left(\frac{\partial \psi}{\partial y}, -\frac{\partial \psi}{\partial x} \right)$$

and then the Navier–Stokes equations effectively become

$$\frac{\partial}{\partial t} \Delta \psi + \frac{\partial(\Delta \psi, \psi)}{\partial(x, y)} = \nu \Delta^2 \psi, \tag{3.3.4}$$

where ν is the kinematic viscosity of the fluid, $\partial(\cdot, \cdot)/\partial(\cdot, \cdot)$ denotes the Jacobian, and Δ is the two-dimensional Laplacian operator. The boundary conditions adopted by Cox (1991) are no slip on u, with constant, uniform suction on v, i.e. for V_1, V_{-1} constants,

$$\begin{aligned} u = 0, v = V_1, \qquad y = h, \\ u = 0, v = -V_{-1}, \qquad y = -h. \end{aligned} \tag{3.3.5}$$

By rescaling (3.3.4) and (3.3.5) he then reduces these to non-dimensional form and then observes there is a similarity solution of form

$$\psi(x, y, t) = x f(y, t),$$

and so equation (3.3.4) may be written (in non-dimensional form)

$$\frac{\partial^3 f}{\partial t \partial y^2} = \frac{1}{R} \frac{\partial^4 f}{\partial y^4} + f \frac{\partial^3 f}{\partial y^3} - \frac{\partial f}{\partial y} \frac{\partial^2 f}{\partial y^2}, \tag{3.3.6}$$

where R is the Reynolds number, in this case defined by

$$R = \frac{(V_1 + V_{-1})h}{2\nu}.$$

Equation (3.3.6) is the Proudman - Johnson equation. The boundary conditions of Cox (1991) are

$$f(1,t) = -1 - \epsilon, \quad \frac{\partial f}{\partial y}(1,t) = 0,$$

$$f(-1,t) = 1 - \epsilon, \quad \frac{\partial f}{\partial y}(-1,t) = 0. \tag{3.3.7}$$

Drazin (1991) relates that equation (3.3.6) was derived by Proudman & Johnson (1962) in their investigation of unsteady flow near a stagnation point. The inviscid version of equation (3.3.6) (i.e. that which follows from Euler's rather than the Navier–Stokes equations) was seen already in (3.2.6). It is worthy of mention that a detailed analysis of the bifurcation picture of the solutions to the Proudman–Johnson equation and the stability of the steady flows is given by Watson *et al.* (1990).

Cox (1991) contains a careful study of the bifurcation behaviour of a solution to (3.3.6), (3.3.7), including a discussion of blow-up and transistion to chaos. He indicates that a numerical solution to (3.3.6), (3.3.7) for R large leads to what looks like blow-up in finite time. However, Cox (1991) also shows that by refining the spatial grid this is seen to be due to numerical inaccuracy. At present, therefore, it would appear to be an open problem whether a solution to (3.3.6) subject to (3.3.7), even with $f = 0$ at $x = \pm 1$, blows up in finite time or not. The numerical evidence provided by Cox (1991) would suggest blow-up is not to be expected.

On a very similar topic, Budd *et al.* (1994) study a variant of (3.3.6), namely

$$\frac{\partial u}{\partial t} - u^2 + \mu v \frac{\partial u}{\partial y} = \frac{\partial^2 u}{\partial y^2} - c,$$

$$\frac{\partial v}{\partial y} = u, \tag{3.3.8}$$

together with the boundary conditions

$$v = 0, \quad \frac{\partial u}{\partial y} = 0, \quad y = 0, 1. \tag{3.3.9}$$

Equations (3.3.8) are defined on the spatial domain $y \in (0,1)$. The above system may be thought of as being motivated by the Navier–Stokes equations with $\nu = 1$, when u, v and p arise from a representation like

$$u \to -xu(y,t), v \to v(y,t), p \to p(y,t) - \frac{1}{2}c(t)x^2.$$

For the Navier–Stokes equations $\mu = 1$ and (3.3.8) is then essentially (3.3.6). Thus, there are two key differences between the work of Budd *et al.* (1994) and that of Childress *et al.* (1989) and Cox (1991). These are that Budd *et al.* (1994) allow a variable μ in front of the vu_y term and they assume a *stress-free boundary condition* on u_y. This would mean prescribing $f_{yy} = 0$

rather than $f_y = 0$ in (3.3.7). Nevertheless, Budd *et al.* (1994) is a very useful piece of work and they write *"there is no evidence that the solutions of the Navier–Stokes equations considered here blow-up (and indeed all the evidence indicates otherwise), the slightly perturbed equations all exhibit blow-up."*

Budd *et al.* (1994) deduce that for $0 \leq \mu < \frac{1}{2}$ blow-up of a solution to (3.3.8), (3.3.9) occurs at $x = 0$ for large enough initial data. For $\frac{1}{2} < \mu < 1$ they also show that there are solutions which blow-up in finite time. When $\mu \geq 1$, however, Budd *et al.* (1994) demonstrate there are solutions to (3.3.8), (3.3.9) which decay to zero as $t \to \infty$.

We might also draw attention to work of Bartuccelli *et al.* (1990) who are interested in turbulence questions for the Navier–Stokes equations. To gain some insight into this complex phenomenon they develop a very interesting analysis for the complex Ginzburg–Landau equation which for certain problems may be derived from the Navier–Stokes equations using multiple scaling arguments. This equation is

$$\frac{\partial A}{\partial t} = RA + (1 + i\nu)\Delta A - (1 + i\mu)A|A|^2.$$

The paper of Bartuccelli *et al.* (1990) derives estimates for the dimension of the attractor and studies stability carefully. Among the various results they achieve are those of finite time blow-up when the spatial domain is greater than or equal to two. This is linked in Bartuccelli *et al.* (1990) to turbulent behaviour.

3.3.1 Self-similar Solutions

Necas *et al.* (1996) remark that Leray (1934) raised the question of the existence of self-similar solutions to the Navier–Stokes equations. Necas *et al.* (1996) investigate the problem of whether solutions to the Navier–Stokes equations

$$\frac{\partial u_i}{\partial t} + u_j \frac{\partial u_i}{\partial x_j} = -\frac{\partial p}{\partial x_i} + \nu \Delta u_i,$$

$$\frac{\partial u_i}{\partial x_i} = 0,$$

(3.3.10)

in $\mathbf{R}^3 \times (t_1, t_2)$ can be of the form

$$u_i(\mathbf{x}, t) = \frac{1}{\sqrt{2a(T - t)}} U_i\left(\frac{\mathbf{x}}{\sqrt{2a(T - t)}}\right).$$

(3.3.11)

A solution like (3.3.11) is of especial interest here since it may well possess a finite-time singularity at $t = T$.

A solution of the form (3.3.11) will from (3.3.10) satisfy the system of partial differential equations

$$-\nu \Delta U_i + aU_i + ay_i \frac{\partial U_i}{\partial x_j} + U_j \frac{\partial U_i}{\partial x_j} = -\frac{\partial P}{\partial x_I},$$

$$\frac{\partial U_i}{\partial x_i} = 0, \qquad (3.3.12)$$

in $\mathbf{R}^3$, where P is a self - similar form of the pressure p. The paper of Necas et al. (1996) establishes the very interesting result that the only solution to (3.3.12) belonging to $L^3(\mathbf{R}^3)$ is $U_i \equiv 0$.

3.3.2 Bénard–Marangoni Convection

Another very interesting field involving the Navier–Stokes equations, in which finite time singularity formation has been observed is in thermal convection with a free surface (interface), as discovered by Thess et al. (1995).

Thess et al. (1995) study three-dimensional thermal convection in a layer in which the free surface is represented mathematically by the plane $z = 0$. The surface tension, σ, is assumed to be a linear function of the temperature field, θ, i.e.

$$\sigma = \sigma_0 - \gamma(\theta - \theta_0). \qquad (3.3.13)$$

The argument of Thess et al. (1995) is that thermal diffusivity is often very much smaller than the kinematic viscosity and can as a first step be neglected. This also leads to consideration of infinite Prandtl number flow. Thus the partial differential equations governing the convection process of Thess et al. (1995) are

$$-\frac{\partial p}{\partial x_i} + \mu \Delta u_i = 0,$$

$$\frac{\partial u_i}{\partial x_i} = 0, \qquad (3.3.14)$$

$$\frac{\partial \theta}{\partial t} + u_i \frac{\partial \theta}{\partial x_i} = 0,$$

where μ is the dynamic viscosity, p is the pressure, and $\mathbf{u}$ is velocity.

Variations in surface tension due to temperature induce velocity gradients at the free surface. To see how these enter the mathematical picture we observe that the boundary condition at the free surface assumes the form

$$t^i = t^{ij} n_j = \sigma b^\alpha_\alpha n^i + a^{\alpha\beta} x^i_{;\alpha} \sigma_{;\beta}, \qquad (3.3.15)$$

where t^i, t^{ij}, n^i denote the stress vector, stress tensor, and unit outward normal, defined here at $z = 0$. The forms b^α_α and $a^{\alpha\beta}$ are twice the mean curvature and the first fundamental form of the surface. For a linear viscous fluid the stress tensor is given by

$$t_{ij} = -p\delta_{ij} + 2\mu d_{ij}, \qquad (3.3.16)$$

where p, μ are the pressure and dynamic viscosity, and $d_{ij} = \frac{1}{2}(u_{i,j} + u_{j,i})$ is the symmetric part of the velocity gradient. For $i = 1, 2$ we employ (3.3.13) and (3.3.16) in (3.3.15) so that the equations

$$t^1 = a^{11}\sigma_{;1}, \qquad t^2 = a^{22}\sigma_{;2},$$

yield the boundary conditions of Thess et al. (1995), at $z = 0$,

$$\mu\frac{\partial u}{\partial z} + \gamma\frac{\partial \theta}{\partial x} = 0,$$
$$\mu\frac{\partial v}{\partial z} + \gamma\frac{\partial \theta}{\partial y} = 0, \qquad\qquad (3.3.17)$$
$$w = 0,$$

where $\mathbf{u} = (u, v, w)$.

Thess et al. (1995) argue that due to the conservation structure of equation(3.3.14)$_3$, particles initially in the free surface remain there for all time. This allows them to study the temperature evolution at $z = 0$ according to (3.3.14)$_3$. To make the system determinate they investigate periodic spatial dependences and then equations (3.3.14)$_{1,2}$ and (3.3.17) allow $\mathbf{u}$ to be expressed as a function of θ leading to a single non-linear equation for θ at $z = 0$. This equation has form

$$\frac{\partial \theta}{\partial t} - (H\theta)\frac{\partial \theta}{\partial x} = 0, \qquad\qquad (3.3.18)$$

where H is the Hilbert transform

$$H\theta = \frac{1}{\pi}\,\text{PV}\int_{-\infty}^{\infty}\frac{\theta(s)}{x - s}\,ds\,,$$

PV being the Cauchy principal value.

Thess et al. (1995) solve equation (3.3.18) numerically using a pseudospectral technique. They observe singularity formation in finite time of $\partial\theta/\partial x$ at the point $x = \pi$, and are able to calculate the time of blow-up, and predict the space-time behaviour of θ near $x = \pi$. Although equation (3.3.18) has to be solved numerically, Thess et al. (1995) show that the extension of this equation to a complex equation is analytically solvable. Thus, they are able to obtain a kind of analytical justification for their numerical calculations.

We understand from Dr. Thess that the singular behaviour discussed above is confined to the infinite layer with periodic spatial solutions. For a finite region he has not seen singularity formation.

4. Catastrophic Behaviour
in Other Non-linear Fluid Theories

4.1 Non-existence on Unbounded Domains

In Sect. 3.3 we have already seen that the Navier–Stokes equations describing the flow of an incompressible, linear viscous fluid present a challenge to whether solutions exist globally or not. In order to have a non-linear fluid theory for which questions of existence, regularity, and formation of turbulence can be better understood from a mathematical point of view Ladyzhenskaya (1967, 1968, 1969) suggested several alternative models and supported these with positive analysis answering many questions which were not able to be solved using Navier–Stokes theory.

4.1.1 Ladyzhenskaya's Models

Ladyzhenskaya (1967, 1968, 1969) has presented a variety of models for nonlinear viscous fluid behaviour. We here discuss four of her modifications, now listed as models I–IV.

Model I

$$\frac{\partial v_i}{\partial t} + v_j \frac{\partial v_i}{\partial x_j} = -\frac{\partial p}{\partial x_i} + (\nu_0 + \nu_1 \|\nabla \mathbf{v}\|^2)\Delta v_i,$$

$$\frac{\partial v_i}{\partial x_i} = 0;$$

(4.1.1)

Model II

$$\frac{\partial v_i}{\partial t} + v_j \frac{\partial v_i}{\partial x_j} = -\frac{\partial p}{\partial x_i} - \left(\operatorname{curl}[(\nu_0 + \nu_1 |\operatorname{curl} \mathbf{v}|^2)\operatorname{curl} \mathbf{v}]\right)_i,$$

$$\frac{\partial v_i}{\partial x_i} = 0.$$

(4.1.2)

It is pertinent at this juncture to note that model II was also proposed and studied by Golovkin (1967).

Model III

$$\frac{\partial v_i}{\partial t} + v_j \frac{\partial v_i}{\partial x_j} = -\frac{\partial p}{\partial x_i} + \frac{\partial}{\partial x_j}\left[(\nu_0 + \nu_1|\nabla \mathbf{v}|^2)\frac{\partial v_i}{\partial x_j}\right],$$
$$\frac{\partial v_i}{\partial x_i} = 0. \tag{4.1.3}$$

In the equations above,

$$\|\nabla \mathbf{v}\|^2 = \int_\Omega v_{i,j} v_{i,j} dV, \qquad |\nabla \mathbf{v}|^2 = v_{i,j} v_{i,j},$$

where Ω is the spatial domain occupied by the fluid, ν_0 and ν_1 are positive constants, and the body force has been assumed conservative and hence absorbed in the pressure p.

Model IV

$$\frac{\partial v_i}{\partial t} + v_j \frac{\partial v_i}{\partial x_j} = -\frac{\partial p}{\partial x_i} + \frac{\partial T_{ij}}{\partial x_j},$$
$$\frac{\partial v_i}{\partial x_i} = 0, \tag{4.1.4}$$

where $T_{ij} = T_{ij}(A_{rs})$, T_{ij} is a symmetric tensor, and $A_{rs} = v_{i,j} + v_{j,i}$. In equation $(4.1.4)_1$ the tensor T_{ij} can have various forms, one of relevance here being,

$$T_{ik} A_{ik} \geq \nu A_{ij} A_{ij}\left[1 + \epsilon(A_{pq} A_{pq})^\mu\right], \tag{4.1.5}$$

where ν, ϵ, and μ are positive constants.

Model IV was given by Ladyzhenskaya (1969), p. 193, and may be viewed as a generalization of model III.

To justify the above models Ladyzhenskaya (1968), pp. 84,85, observes that the temperature T is assumed constant in the derivation of the Navier–Stokes equations from the Maxwell–Boltzmann equations. However, in the absence of such an assumption the system of partial differential equations to be solved for the velocity $\mathbf{v}$, pressure p, and temperature T, is,

$$\frac{\partial v_i}{\partial t} + v_j \frac{\partial v_i}{\partial x_j} = -\frac{\partial p}{\partial x_i} + c\frac{\partial}{\partial x_j}\left(\sqrt{T}\frac{\partial v_i}{\partial x_j}\right),$$
$$\frac{\partial v_i}{\partial x_i} = 0, \tag{4.1.6}$$
$$\frac{\partial T}{\partial t} + v_i \frac{\partial T}{\partial x_i} = c_1\frac{\partial}{\partial x_i}\left(\sqrt{T}\frac{\partial T}{\partial x_i}\right) + \frac{1}{2}c\sqrt{T}A_{ij}A_{ij},$$

where c and c_1 are positive constants.

To motivate models I and III, Ladyzhenskaya (1968) observes that when $\Omega = \mathbf{R}^3$ it follows from the maximum principle that

$$
\min_{\mathbf{x} \in \mathbf{R}^3} \sqrt{T(\mathbf{x}, 0)} + \frac{cT_1}{4} \min_{\mathbf{R}^3 \times [0, t_1]} A_{ij} A_{ij}
$$
$$
\leq \sqrt{T(\mathbf{x}, t_1)} \leq \max_{\mathbf{x} \in \mathbf{R}^3} \sqrt{T(\mathbf{x}, 0)} + \frac{cT_1}{4} \max_{\mathbf{R}^3 \times [0, t_1]} A_{ij} A_{ij} ,
$$

(4.1.7)

for t_1 a fixed positive time. On this basis she then argues that (4.1.7) suggest that $c\sqrt{T}$ may be replaced in (4.1.6)$_1$ by $\nu_0 + \nu_1 v_{i,j} v_{i,j}$, or by $\nu_0 + \nu_1 \|\nabla \mathbf{v}\|^2$.

In this book we are interested in the blow-up and non-existence of solutions. Consider, therefore, the *backward in time* problem in the context of model I, model II, or model III.

One may ask why one should consider a fluid problem for values of time in the past. There are many situations where one needs to extrapolate a model to previous times in order to glean information which will be useful for future prediction. For instance, history matching properties of rocks, etc., in an underground oil reservoir is a well known area. Studies of the Earth's magnetic field require investigation of the history of the field. A theoretician could always guess at what conditions were like many years ago and compute forward in time, but to match conditions exactly today may not be easy and hence the desire to study the backward evolution of the solution to a model from known current conditions. Many instances of deriving stabilizing continuous dependence estimates for backward in time problems are considered in the book by Ames & Straughan (1997).

If Ω is bounded and zero boundary data are employed Straughan (1976) has shown that solutions to (4.1.1), (4.1.2), or (4.1.3) do not exist globally in the backward in time problem. Furthermore, Ames *et al.* (1989) investigate the above backward in time problems when Ω is either $\mathbf{R}^2$ or $\mathbf{R}^3$. In the interests of clarity, and to facilitate the calculations which follow we now rewrite models I–III backward in time by reversing time in the relevant partial differential equations so that $t > 0$, and we then study the transformed partial differential equations forward in time. In this case the velocity and pressure fields (v_i, p) satisfy the coupled system of partial differential equations

Model I

$$
\frac{\partial v_i}{\partial t} = v_j \frac{\partial v_i}{\partial x_j} + \frac{\partial p}{\partial x_i} - (\nu_0 + \nu_1 \|\nabla \mathbf{v}\|^2) \Delta v_i,
$$
$$
\frac{\partial v_i}{\partial x_i} = 0;
$$

(4.1.8)

Model II

$$\frac{\partial v_i}{\partial t} = v_j \frac{\partial v_i}{\partial x_j} + \frac{\partial p}{\partial x_i} + \left(\text{curl} \left[(\nu_0 + \nu_1 |\text{curl } \mathbf{v}|^2) \text{curl } \mathbf{v} \right] \right)_i,$$

$$\frac{\partial v_i}{\partial x_i} = 0;$$

(4.1.9)

Model III

$$\frac{\partial v_i}{\partial t} = v_j \frac{\partial v_i}{\partial x_j} + \frac{\partial p}{\partial x_i} - \frac{\partial}{\partial x_j} \left[(\nu_0 + \nu_1 |\nabla \mathbf{v}|^2) \frac{\partial v_i}{\partial x_j} \right],$$

$$\frac{\partial v_i}{\partial x_i} = 0;$$

(4.1.10)

The solution to each of these systems of partial differential equations will be accompanied by initial data

$$\mathbf{v}(\mathbf{x}, 0) = \mathbf{v}_0(\mathbf{x}), \qquad \mathbf{x} \in \Omega. \tag{4.1.11}$$

The proofs of global non-existence in Straughan (1976) rely on Poincaré's inequality in the form

$$\|\nabla \mathbf{v}\| \geq \lambda_1 \|\mathbf{v}\|^2, \qquad \lambda_1 > 0.$$

In particular, these proofs break down when the region Ω is unbounded in all spatial directions and an entirely different analysis is necessarily adopted by Ames *et al.* (1989).

Breakdown also occurs in the corresponding problem for model IV, i.e. that governed by the system of partial differential equations (4.1.4), when $t < 0$. In this case the velocity and pressure $(\mathbf{v}, p)$ satisfies model IV backward in time which we write by reversing the sign of t in (4.1.4) as

$$\frac{\partial v_i}{\partial t} = v_j \frac{\partial v_i}{\partial x_j} + \frac{\partial p}{\partial x_i} - \frac{\partial T_{ij}}{\partial x_j},$$

$$\frac{\partial v_i}{\partial x_i} = 0,$$

(4.1.12)

where the stress tensor T_{ij} is subject to inequality (4.1.5).

If now $\Omega \subset \mathbf{R}^3$ is bounded and homogeneous boundary data are given, i.e.

$$v_i(\mathbf{x}, 0) = 0 \qquad \text{on } \Gamma, \tag{4.1.13}$$

where Γ is the boundary of Ω, then we show below that the solution to (4.1.12), (4.1.13) and (4.1.5) cannot exist for all time.

To establish global non-existence we commence as follows. Multiply equation $(4.1.12)_1$ by v_i and integrate over Ω to find with the aid of the divergence theorem and $(4.1.13)$,

$$\frac{d}{dt}\|\mathbf{v}\|^2 = 2\int_\Omega T_{ij}v_{i,j}\,dx$$

$$= \int_\Omega T_{ij}A_{ij}\,dx$$

$$\geq \nu\left[\int_\Omega A_{ij}A_{ij}\,dx + \epsilon\int_\Omega (A_{ij}A_{ij})^{1+\mu}dx\right], \qquad (4.1.14)$$

where in the last line $(4.1.5)$ has been employed. Note that

$$\frac{1}{2}\int_\Omega A_{ij}A_{ij}\,dx = \frac{1}{2}\int_\Omega (v_{i,j}+v_{j,i})(v_{i,j}+v_{j,i})\,dx$$

$$= \int_\Omega v_{i,j}v_{i,j}\,dx, \qquad (4.1.15)$$

as can be seen using the boundary conditions and the solenoidal nature of $\mathbf{v}$.
From Hölder's inequality we easily find

$$\int_\Omega (A_{ij}A_{ij})^{1+\mu}dx \geq c\left(\int_\Omega A_{ij}A_{ij}\,dx\right)^{1+\mu}, \qquad (4.1.16)$$

where c is a constant depending on Ω and c becomes vanishingly small the larger Ω becomes.

Upon employment of $(4.1.15)$ and $(4.1.16)$ in $(4.1.14)$ we deduce

$$\frac{d}{dt}\|\mathbf{v}\|^2 \geq 2\nu\int_\Omega v_{i,j}v_{i,j}\,dx + \epsilon\nu c2^{1+\mu}\left(\int_\Omega v_{i,j}v_{i,j}\,dx\right)^{1+\mu}. \qquad (4.1.17)$$

Next, use inequality $(4.1.11)$ in $(4.1.17)$ to find

$$\frac{dJ}{dt} \geq 2\nu\lambda_1 J + (2\lambda_1)^{1+\mu}\nu\epsilon c J^{1+\mu}, \qquad (4.1.18)$$

where $J(t)$ is defined as

$$J(t) = \|\mathbf{v}\|^2.$$

Inequality $(4.1.18)$ is of Bernoulli type and so integrates to yield:

$$J^\mu(t) \geq \frac{J^\mu(0)e^{2\nu\lambda_1\mu t}}{1 - 2^\mu\lambda_1^\mu J^\mu(0)c\epsilon(e^{2\nu\lambda_1\mu t}-1)}. \qquad (4.1.19)$$

The right hand side of this expression blows up in a finite time and so the solution to $(4.1.12)$ cannot exist beyond this time. An upper bound for the interval of existence is thus found from $(4.1.19)$ to be

$$T = \frac{1}{2\nu\lambda_1\mu} \log\left(1 + \frac{1}{(2\lambda_1\|\mathbf{v}_0\|^2)^\mu \epsilon c}\right).$$

Similar finite time non-existence results are given in Straughan (1976) for models I–III. Again, we point out that the proofs depend on Ω being bounded. For instance, the above proof requires (4.1.11) and (4.1.16), which do not hold for Ω unbounded in all directions.

We believe it is worth pointing out that if we consider instead of model I, model II, model III, or model IV, Ladyzhenskaya's original model (4.1.6), but for the backward in time situation, i.e. assume $t < 0$, with $v_i = 0$ and $\partial T/\partial n = 0$ on the boundary (thermally insulated) then the foregoing non-existence arguments do not work and a different outcome is suggested. In this case the following "energy-like" identities may be established,

$$\frac{d}{dt}\|\mathbf{v}\|^2 = c\int_\Omega \sqrt{T}|\nabla\mathbf{v}|^2 dx,$$

$$\frac{d}{dt}\|T\|^2 = c_1\int_\Omega \sqrt{T}|\nabla T|^2 dx - \frac{1}{2}c\int_\Omega T^{3/2}A_{ij}A_{ij} dx.$$

The stress power term in the partial differential equation $(4.1.6)_3$ gives rise to the cubic (negative) term in the second of the above equations and this prevents the non-existence arguments from applying. It is possible that the solution exists globally. Ladyzhenskaya (1968) writes that she has not established global existence for a solution to (4.1.6), but it would seem that the properties of this system are likely to be of much interest.

We now turn to an analysis of each of models I-III, backward in time, i.e. models governed by the system of partial differential equations (4.1.8), (4.1.9), or (4.1.10), when Ω is either $\mathbf{R}^2$ or $\mathbf{R}^3$. Thus, we are providing an exposition of the work covered in Ames *et al.* (1989).

4.1.2 Global Non-existence Backward in Time for Model I, When the Spatial Domain Is $\mathbf{R}^2$

In the interests of clarity we here rewrite the equations for model I with $t > 0$ (although it is to be understood we are considering the final value problem):

$$\frac{\partial v_i}{\partial t} = v_j\frac{\partial v_i}{\partial x_j} + \frac{\partial p}{\partial x_i} - (\nu_0 + \nu_1\|\nabla\mathbf{v}\|^2)\Delta v_i,$$

$$\frac{\partial v_i}{\partial x_i} = 0. \tag{4.1.20}$$

The work of Ames *et al.* (1989) requires a separate treatment for the cases $\Omega = \mathbf{R}^2$ and $\Omega = \mathbf{R}^3$. In both cases the solution and its derivatives are assumed to vanish in the correct manner as $|\mathbf{x}| \to \infty$ in order that the required functions are members of $L^2(\Omega)$. The initial data are given in the form,

$$\mathbf{v}(\mathbf{x}, 0) = \mathbf{v}^0(\mathbf{x}), \qquad \mathbf{x} \in \Omega. \tag{4.1.21}$$

Suppose now $\Omega = \mathbf{R}^2$. The non-existence proof begins by defining $J(t)$ by

$$J(t) = \|\mathbf{v}\|^2.$$

A standard calculation substituting for $v_{i,t}$ from (4.1.20) and integration by parts shows

$$J'(t) = 2\nu_0 I + 2\nu_1 I^2, \tag{4.1.22}$$

where $I(t)$ is given by

$$I(t) = \|\nabla \mathbf{v}\|^2. \tag{4.1.23}$$

Subsequent calculations require the derivative of I, and again employing (4.1.20) we may show that

$$\begin{aligned}
\frac{dI}{dt} &= -2(\Delta \mathbf{v}, \mathbf{v}_t), \\
&= 2\big(\Delta \mathbf{v}, (\nu_0 + \nu_1 \|\nabla \mathbf{v}\|^2)\Delta \mathbf{v} - \nabla p - (\mathbf{v} \cdot \nabla)\mathbf{v}\big), \\
&= 2\nu_0\|\Delta \mathbf{v}\|^2 + 2\nu_1 I \|\Delta \mathbf{v}\|^2, \tag{4.1.24}
\end{aligned}$$

where in deriving (4.1.24) we have used the relations

$$(\Delta \mathbf{v}, \nabla p) = 0$$

and

$$-((\mathbf{v} \cdot \nabla)\mathbf{v}, \Delta \mathbf{v}) = \int_\Omega \frac{\partial v_i}{\partial x_j} \frac{\partial v_i}{\partial x_k} \frac{\partial v_j}{\partial x_k}\, dx = 0.$$

The last fact follows by expanding the integrand in the middle term and using the expression

$$\frac{\partial v_1}{\partial x_1} + \frac{\partial v_2}{\partial x_2} = 0.$$

The pressure term also vanishes thanks to $v_{i,i} = 0$.

In the ensuing analysis we utilize the inequality

$$\|\Delta \mathbf{v}\|^2 \geq \frac{\|\nabla \mathbf{v}\|^4}{\|\mathbf{v}\|^2}, \tag{4.1.25}$$

which follows since

$$\|\nabla \mathbf{v}\|^2 = -(\mathbf{v}, \Delta \mathbf{v}) \leq \|\mathbf{v}\|\,\|\Delta \mathbf{v}\|,$$

where in the last stage the Cauchy–Schwarz inequality has been employed. We also have recourse to use the derivative of I^2 and this is seen to be

$$\begin{aligned}
\frac{d}{dt}\|\nabla \mathbf{v}\|^4 &= 2\|\nabla \mathbf{v}\|^2 \frac{dI}{dt}, \\
&= 4\nu_0\|\nabla \mathbf{v}\|^2\|\Delta \mathbf{v}\|^2 + 4\nu_1\|\nabla \mathbf{v}\|^4\|\Delta \mathbf{v}\|^2, \tag{4.1.26}
\end{aligned}$$

with the help of (4.1.24).

The next step is to differentiate (4.1.22) to obtain, using (4.1.24) and (4.1.26),

$$J'' = 4\nu_0^2\|\Delta\mathbf{v}\|^2 + 12\nu_0\nu_1\|\nabla\mathbf{v}\|^2\|\Delta\mathbf{v}\|^2 + 8\nu_1^2\|\nabla\mathbf{v}\|^4\|\Delta\mathbf{v}\|^2,$$

$$\geq \left(4\nu_0^2 + 12\nu_0\nu_1\|\nabla\mathbf{v}\|^2 + 8\nu_1^2\|\nabla\mathbf{v}\|^4\right)\frac{\|\nabla\mathbf{v}\|^4}{\|\mathbf{v}\|^2}, \qquad (4.1.27)$$

where (4.1.25) has been employed.

We next observe

$$(J')^2 = 4\left(\nu_0\|\nabla\mathbf{v}\|^2 + \nu_1\|\nabla\mathbf{v}\|^4\right)^2$$

and so by multiplying (4.1.27) by J and rearranging we derive

$$JJ'' \geq -2\nu_0^2\|\nabla\mathbf{v}\|^4 + \frac{3}{2}(J')^2 + 2\nu_1^2\|\nabla\mathbf{v}\|^8. \qquad (4.1.28)$$

The last term is discarded and from (4.1.22) we see that

$$-2\|\nabla\mathbf{v}\|^4 = -\frac{J'}{\nu_1} + 2\frac{\nu_0}{\nu_1}I,$$

and hence from (4.1.28) we find

$$JJ'' - \frac{3}{2}(J')^2 \geq -\frac{\nu_0^2}{\nu_1}J'. \qquad (4.1.29)$$

To employ the method of concavity we observe that from (4.1.22)

$$J(t) \geq J(0),$$

which used in (4.1.29) yields

$$JJ'' - \frac{3}{2}(J')^2 \geq -\frac{\nu_0^2}{\nu_1 J(0)}JJ'. \qquad (4.1.30)$$

If we multiply by $-\frac{1}{2}J^{-5/2}$, (4.1.30) may be written

$$(J^{-1/2})'' \leq -\frac{\nu_0^2}{\nu_1 J(0)}(J^{-1/2})'. \qquad (4.1.31)$$

This inequality is now integrated once in t and then the resulting first-order differential inequality is integrated with the aid of an integrating factor to obtain, after some rearrangement,

$$J^{1/2}(t) \geq \frac{J^{1/2}(0)e^{kt}}{1 - \left[(\nu_1 J'(0)/2\nu_0^2) - 1\right](e^{kt} - 1)}, \qquad (4.1.32)$$

where

$$k = \frac{\nu_0^2}{\nu_1 J(0)}. \tag{4.1.33}$$

Provided

$$J'(0) > \frac{2\nu_0^2}{\nu_1}, \tag{4.1.34}$$

the right hand side of (4.1.32) blows up in a finite time, T, with

$$T = \frac{\nu_1 J(0)}{\nu_0^2} \log \left(\frac{J'(0)}{J'(0) - 2\nu_0^2/\nu_1} \right), \tag{4.1.35}$$

and so the solution cannot exist beyond this time.

We may instead proceed directly from (4.1.29). If this inequality is multiplied by $-\frac{1}{2}J^{-5/2}$, we see that

$$(J^{-1/2})'' \le \frac{\nu_0^2}{2\nu_1} \frac{J'}{J^{5/2}} = -\frac{\nu_0^2}{3\nu_1}(J^{-3/2})'.$$

Upon integration one obtains

$$(J^{-1/2})' \le -\frac{\nu_0^2}{3\nu_1 J^{3/2}} + \frac{\nu_0^2}{3\nu_1 J^{3/2}(0)} - \frac{J'(0)}{2J^{3/2}(0)}.$$

Evaluating the derivative and rearranging,

$$J' \ge \frac{2\nu_0^2}{3\nu_1} + \frac{1}{J^{3/2}(0)} \left[J'(0) - \frac{2\nu_0^2}{3\nu_1} \right] J^{3/2}. \tag{4.1.36}$$

Nonexistence now follows by a variation of the convergent integral method. If

$$J'(0) > \frac{2\nu_0^2}{3\nu_1}, \tag{4.1.37}$$

then we separate variables in (4.1.35) to derive

$$t \le \int_{J(0)}^\infty \frac{3\nu_1 J^{3/2}(0) ds}{2\nu_0^2 J^{3/2}(0) + (3\nu_1 J'(0) - 2\nu_0^2)s^{3/2}} < \infty. \tag{4.1.38}$$

The integral in (4.1.38) is the analogous estimate to (4.1.35) when derived by the above technique.

In addition to the above methods Ames *et al.* (1989) also employ a third technique which requires no initial data restriction. In that case they show that an estimate for the maximum interval of existence is

$$T = \frac{J(0)}{2\nu_0 I(0)} \log \left[1 + \frac{\nu_0}{\nu_1 I(0)} \right].$$

They also prove that an upper bound for the growth of $J(t)$ may be found if the blow up time, t^*, is known. In particular, they demonstrate that either

$$J(t) \leq \frac{\nu_0 J(0)}{\nu_1 I(0)\left(\exp\left[2\nu_0 I(0)(t^* - t)J(0)^{-1}\right] - 1\right)} \, ,$$

or

$$J(t) \leq \left(\frac{1 - e^{-kt^*}}{e^{-kt} - e^{-kt^*}}\right)^2 J(0),$$

where k is given by (4.1.33).

4.1.3 Global Nonexistence Backward in Time for Model I, When the Spatial Domain Is $\mathbf{R}^3$

The initial value problem (4.1.20), (4.1.21) is now investigated when the spatial domain $\Omega = \mathbf{R}^3$.

Equation (4.1.22) still holds. However, equation (4.1.24) does not since the term

$$\left(\Delta \mathbf{v}, (\mathbf{v} \cdot \nabla)\mathbf{v}\right)$$

is not necessarily zero. Therefore, equation (4.1.24) must be replaced by

$$\frac{dI}{dt} = 2\nu_0 \|\Delta \mathbf{v}\|^2 + 2\nu_1 I \|\Delta \mathbf{v}\|^2 - 2\left(\Delta \mathbf{v}, (\mathbf{v} \cdot \nabla)\mathbf{v}\right). \tag{4.1.39}$$

There are several ways to estimate the last term. We may, for example, write

$$\left(\Delta \mathbf{v}, (\mathbf{v} \cdot \nabla)\mathbf{v}\right) \leq \sup_{\mathbf{x} \in \Omega} |\mathbf{v}| \, \|\nabla \mathbf{v}\| \, \|\Delta \mathbf{v}\|,$$

and then employ Xie's (1991) inequality

$$\sup_{\mathbf{x} \in \Omega} |\mathbf{v}| \leq \frac{1}{\sqrt{2\pi}} \|\nabla \mathbf{v}\|^{1/2} \|\Delta \mathbf{v}\|^{1/2},$$

where the constant is optimal. (The use of this inequality requires $|\mathbf{v}| \to 0$ as $|\mathbf{x}| \to \infty$, which is implicit in the calculations.) Then,

$$\left(\Delta \mathbf{v}, (\mathbf{v} \cdot \nabla)\mathbf{v}\right) \leq \frac{1}{\sqrt{2\pi}} \|\nabla \mathbf{v}\|^{3/2} \|\Delta \mathbf{v}\|^{3/2},$$

$$\leq \frac{1}{\sqrt{2\pi}} \|\mathbf{v}\|^{1/2} \|\nabla \mathbf{v}\|^{1/2} \|\Delta \mathbf{v}\|^2, \tag{4.1.40}$$

where inequality (4.1.25) has been employed in the last line.

Inequality (4.1.40) is now used in equation (4.1.39), to see that

$$\frac{dI}{dt} \geq 2(\nu_0 + \nu_1 I) \|\Delta \mathbf{v}\|^2 - \sqrt{\frac{2}{\pi}} I^{1/4} J^{1/4} \|\Delta \mathbf{v}\|^2.$$

Ames *et al.* (1989) rewrite this inequality as

$$\frac{dI}{dt} \geq (1+\alpha)(\nu_0 + \nu_1 I)\|\Delta\mathbf{v}\|^2$$

$$+ \left[(1-\alpha)(\nu_0 + \nu_1 I)\|\Delta\mathbf{v}\|^2 - \sqrt{\frac{2}{\pi}}I^{1/4}J^{1/4}\right]\|\Delta\mathbf{v}\|^2, \tag{4.1.41}$$

where α is an arbitrary number in $(0,1)$.

Ames *et al.* (1989) further show that provided

$$(1-\alpha)\left(\nu_0 + \nu_1 I(0)\right) > \sqrt{\frac{2}{\pi}}I^{1/4}(0)J^{1/4}(0), \tag{4.1.42}$$

and,

$$\frac{\nu_1}{\nu_0}\left(\frac{1+3\alpha}{3+\alpha}\right)\frac{(I(0))^{2/(3+\alpha)}}{(J(0))^{(1+\alpha)/(3+\alpha)}}$$

$$\geq \left[\frac{1+3\alpha}{2^{5/2}\pi^{1/2}\nu_0(1-\alpha^2)}\right]^{4(1+\alpha)/(3+\alpha)}, \tag{4.1.43}$$

then the quantity

$$(1-\alpha)\left(\nu_0 + \nu_1 I(t)\right) - \sqrt{\frac{2}{\pi}}(I(t))^{1/4}(J(t))^{1/4},$$

is positive for all t for which the solution exists. It then follows from (6.3.41) that, also for all t for which the solution exists,

$$\frac{dI}{dt} \geq (1+\alpha)(\nu_0 + \nu_1 I)\|\Delta\mathbf{v}\|^2$$

$$\geq (1+\alpha)(\nu_0 + \nu_1 I)\frac{I^2}{J}, \tag{4.1.44}$$

with the aid of (4.1.25). There are several ways in which one may proceed from this. For example, we may compute J'' from (4.1.22),

$$J'' = (2\nu_0 + 4\nu_1 I)I'$$

$$\geq 2(1+\alpha)(\nu_0 I + 2\nu_1 I^2)\frac{(\nu_0 I + \nu_1 I^2)}{J}, \tag{4.1.45}$$

where we have substituted for I' using (4.1.44). Since from (4.1.22),

$$(J')^2 = 4(\nu_0 I + \nu_1 I^2)^2,$$

we find from (4.1.45), that

$$JJ'' \geq (1+\alpha)(J')^2 - \nu_0(1+\alpha)IJ'. \tag{4.1.46}$$

To derive non-existence from (4.1.46) via the concavity method we note that from (4.1.22),

$$-I = \frac{\nu_0}{\nu_1} - \frac{J'}{2\nu_1 I}. \tag{4.1.47}$$

Furthermore, from (4.1.44),

$$I(t) \geq I(0),$$

and then we may deduce from (4.1.47) (noting $J' \geq 0$),

$$-I \geq \frac{\nu_0}{\nu_1} - \frac{J'}{2\nu_1 I(0)}. \tag{4.1.48}$$

Therefore, (4.1.46) leads to

$$JJ'' - (1+\alpha)\left(1 - \frac{\nu_0}{2\nu_1 I(0)}\right)(J')^2 \geq \frac{\nu_0^2}{\nu_1}(1+\alpha)J'. \tag{4.1.49}$$

Define

$$\zeta = (1+\alpha)\left(1 - \frac{\nu_0}{2\nu_1 I(0)}\right)$$

and suppose $\zeta > 1$. This, of course, means we impose a further data restriction, namely,

$$I(0) > \frac{\nu_0(1+\alpha)}{2\alpha\nu_1}. \tag{4.1.50}$$

Assuming this, we note that since $J' \geq 0$, (4.1.49) may be multiplied by $-\zeta J^{-\zeta-1}$ to obtain

$$\frac{d^2}{dt^2}(J^{-\zeta}) \leq 0,$$

and so by integration,

$$J^\zeta(t) \geq \frac{J^\zeta(0)}{1 - \zeta(J'(0)/J(0))t}. \tag{4.1.51}$$

From (4.1.51) it is easily seen that $\mathbf{v}$ cannot exist globally, and an upper bound for the existence interval is

$$T = \frac{2\nu_1 I(0)J(0)}{J'(0)(1+\alpha)(2\nu_1 I(0) - \nu_0)}.$$

Alternatively, one may proceed from (4.1.49) to find

$$\frac{d^2}{dt^2}(J^{-\zeta}) \leq \frac{\zeta\nu_0^2(1+\alpha)}{\nu_1(1+\zeta)}\frac{d}{dt}(J^{-\zeta-1}).$$

Upon integration and rearrangement we derive

$$\frac{dJ}{dt} \geq c_1 J^{\zeta+1} - c_2, \tag{4.1.52}$$

where

$$c_1 = \zeta J^{-\zeta-1}(0)\left[J'(0) - \frac{\nu_0^2(1+\alpha)}{\nu_1(1+\zeta)}\right],$$

$$c_2 = \frac{\nu_0^2(1+\alpha)}{\nu_1(1+\zeta)}.$$

After separation of variables and integration in (4.1.52) the convergent integral method yields global non-existence with an upper bound for the existence interval

$$T = \int_{J(0)}^{\infty} \frac{ds}{c_1 s^{1+\zeta} - c_2}.$$

A third method is provided by Ames *et al.* (1989). From (4.1.44) and (4.1.22) we see that

$$I' \geq \left(\frac{1+\alpha}{2}\right)\frac{J'I}{J}.$$

Hence,

$$\frac{I'}{I} \geq \left(\frac{1+\alpha}{2}\right)\frac{J'}{J}.$$

After integration this leads to

$$\log\frac{I}{I(0)} \geq \log\left(\frac{J}{J(0)}\right)^{(1+\alpha)/2},$$

or,

$$I(t) \geq \frac{I(0)}{[J(0)]^{(1+\alpha)/2}}[J(t)]^{(1+\alpha)/2}. \qquad (4.1.53)$$

This may be used directly in (4.1.22) to find

$$J' > k_1 J^{(1+\alpha)/2} + k_2 J^{1+\alpha}, \qquad (4.1.54)$$

where k_1 and k_2 are positive constants easily calculated from (4.1.22) and (4.1.53). The convergent integral method applied to (4.1.54) yields an upper bound for the interval of existence as

$$T = \int_{J(0)}^{\infty} \frac{ds}{k_1 s^{(1+\alpha)/2} + k_2 s^{1+\alpha}}.$$

We point out that various other bounds and continuous dependence results for Model I (and Models II and III) are derived in Ames *et al.* (1989).

4.1.4 Exponential Growth for Model II, Backward in Time

To be precise about what we are doing we rewrite the equations for model II, backward in time. Again, we let $t > 0$, although it is understood it is the backward in time problem which is under consideration.

$$\frac{\partial v_i}{\partial t} = v_j \frac{\partial v_i}{\partial x_j} + \frac{\partial p}{\partial x_i} + \left(\text{curl} \left[(\nu_0 + \nu_1 |\text{curl } \mathbf{v}|^2) \text{curl } \mathbf{v} \right] \right)_i,$$

$$\frac{\partial v_i}{\partial x_i} = 0.$$

$$(4.1.55)$$

The work of Ames *et al.* (1989) again considers the problems in $\mathbf{R}^2$ and $\mathbf{R}^3$ separately. We here only deal with the case $\Omega = \mathbf{R}^2$. Initial data are again specified,

$$\mathbf{v}(\mathbf{x}, 0) = \mathbf{v}^0(\mathbf{x}), \qquad \mathbf{x} \in \Omega. \tag{4.1.56}$$

It is convenient to rewrite $(4.1.55)_1$ as

$$\frac{\partial \mathbf{v}}{\partial t} = (\mathbf{v} \cdot \nabla)\mathbf{v} + \nabla p + L\mathbf{v} + M\mathbf{v}, \tag{4.1.57}$$

where the operators $L\mathbf{v}$ and $M\mathbf{v}$ are defined by

$$L\mathbf{v} = \nu_0 \, \text{curl curl } \mathbf{v}, \tag{4.1.58}$$

$$M\mathbf{v} = \nu_1 \, \text{curl} \left(|\text{curl } \mathbf{v}|^2 \text{curl } \mathbf{v} \right). \tag{4.1.59}$$

The analysis again commences with $I(t)$ defined by

$$I(t) = \|\mathbf{v}(t)\|^2. \tag{4.1.60}$$

Then, multiplying (4.1.57) by $\mathbf{v}$ and integrating, using $(4.1.55)_2$, we find

$$\frac{dI}{dt} = 2\nu_0 \|\text{curl } \mathbf{v}\|^2 + 2\nu_1 \int_\Omega |\text{curl } \mathbf{v}|^4 dx. \tag{4.1.61}$$

Next, define $\Phi(t)$ by

$$\Phi = 2\nu_0 \|\text{curl } \mathbf{v}\|^2 + \nu_1 \int_\Omega |\text{curl } \mathbf{v}|^4 dx. \tag{4.1.62}$$

Note that from (4.1.61) it follows immediately that

$$\frac{dI}{dt} \geq \Phi(t). \tag{4.1.63}$$

We use the fact that for functions of the required integrability

$$(\mathbf{a}, \text{curl } \mathbf{b}) = (\text{curl } \mathbf{a}, \mathbf{b}),$$

and find upon differentiation of (4.1.62),

$$\frac{d\Phi}{dt} = 4\nu_0(\operatorname{curl}\operatorname{curl}\mathbf{v}, \mathbf{v}_t) + 4\nu_1 \int_\Omega \operatorname{curl}\left(|\operatorname{curl}\mathbf{v}|^2\operatorname{curl}\mathbf{v}\right) \cdot \mathbf{v}_t dx,$$

$$= 4(\mathbf{v}_t, L\mathbf{v} + M\mathbf{v}),$$

$$= 4\|L\mathbf{v} + M\mathbf{v}\|^2 + 4\left(L\mathbf{v} + M\mathbf{v}, \nabla p + (\mathbf{v} \cdot \nabla)\mathbf{v}\right). \qquad (4.1.64)$$

Since $L\mathbf{v} = -\nu_0\Delta\mathbf{v}$, the term

$$4\left(L\mathbf{v}, \nabla p + (\mathbf{v} \cdot \nabla)\mathbf{v}\right)$$

is zero by the same reasoning as that employed in the analysis for model I. Furthermore,

$$(M\mathbf{v}, \nabla p) = \nu_1(|\operatorname{curl}\mathbf{v}|^2\operatorname{curl}\mathbf{v}, \operatorname{curl}\nabla p) = 0.$$

To handle the term

$$(M\mathbf{v}, (\mathbf{v} \cdot \nabla)\mathbf{v})$$

we recall that $\Omega = \mathbf{R}^2$, and $v_{1,1} + v_{2,2} = 0$. Put $\omega_i = (\operatorname{curl}\mathbf{v})_i$, and then $\omega^2 = (v_{1,2} - v_{2,1})^2$. Using standard tensor calculus we find

$$(M\mathbf{v}, (\mathbf{v} \cdot \nabla)\mathbf{v}) = \int_\Omega \epsilon_{ijk}\frac{\partial}{\partial x_j}\left(\omega^2\epsilon_{krs}\frac{\partial v_s}{\partial x_r}\right)v_m\frac{\partial v_i}{\partial x_m}\,dx,$$

$$= I_1 + I_2,$$

where I_1 and I_2 are defined by

$$I_1 = \int_\Omega \omega^2\left(\frac{\partial v_i}{\partial x_j} - \frac{\partial v_j}{\partial x_i}\right)\frac{\partial v_m}{\partial x_j}\frac{\partial v_i}{\partial x_m}\,dx,$$

$$I_2 = \int_\Omega \omega^2\left(\frac{\partial v_i}{\partial x_j} - \frac{\partial v_j}{\partial x_i}\right)v_m\frac{\partial^2 v_i}{\partial x_j\partial x_m}\,dx.$$

Recalling we are in two dimensions, the integrand of I_1 is easily seen to be

$$\left(\frac{\partial v_1}{\partial x_2} - \frac{\partial v_2}{\partial x_1}\right)^4\left(\frac{\partial v_1}{\partial x_1} + \frac{\partial v_2}{\partial x_2}\right) = 0.$$

Also, expanding I_2 we obtain

$$I_2 = \int_\Omega \omega^2\left(\frac{\partial v_1}{\partial x_2} - \frac{\partial v_2}{\partial x_1}\right)\left[v_1\frac{\partial}{\partial x_1}\left(\frac{\partial v_1}{\partial x_2} - \frac{\partial v_2}{\partial x_1}\right) + v_2\frac{\partial}{\partial x_2}\left(\frac{\partial v_1}{\partial x_2} - \frac{\partial v_2}{\partial x_1}\right)\right]dx,$$

$$= \frac{1}{4}\int_\Omega \left(v_1\frac{\partial}{\partial x_1}\omega^4 + v_2\frac{\partial}{\partial x_2}\omega^4\right)dx,$$

$$= -\frac{1}{4}\int_\Omega \omega^4\frac{\partial v_i}{\partial x_i}\,dx,$$

$$= 0.$$

Then (4.1.64) reduces to

$$\frac{d\Phi}{dt} = 4\|L\mathbf{v} + M\mathbf{v}\|^2.$$
(4.1.65)

From (4.1.62),

$$I' = 2\nu_0\|\text{curl}\,\mathbf{v}\|^2 + 2\nu_1 \int_\Omega |\text{curl}\,\mathbf{v}|^4 dx,$$

$$= 2(\mathbf{v}, L\mathbf{v} + M\mathbf{v}),$$

$$\leq 2\|\mathbf{v}\|\,\|L\mathbf{v} + M\mathbf{v}\|,$$

by the Cauchy-Schwarz inequality. Thus,

$$\|L\mathbf{v} + M\mathbf{v}\|^2 \geq \frac{(I')^2}{4\|\mathbf{v}\|^2}.$$
(4.1.66)

Use of (4.1.66) in (4.1.65) leads to

$$\Phi' \geq \frac{(I')^2}{I},$$

and then taking account of (4.1.63) we find

$$\frac{\Phi'}{\Phi} \geq \frac{I'}{I}.$$
(4.1.67)

Thus,

$$(\log \Phi)' \geq (\log I)',$$

and so

$$\frac{\Phi(t)}{\Phi(0)} \geq \frac{I(t)}{I(0)}.$$

Therefore, use of this in conjunction with (4.1.63) allows us to see

$$I' \geq \Phi \geq I \frac{\Phi(0)}{I(0)}.$$

Upon a further integration we deduce

$$I(t) \geq I(0) \exp\left[\frac{\Phi(0)t}{I(0)}\right].$$
(4.1.68)

It follows from (4.1.68) that the solution to model II cannot remain bounded for all time.

When $\Omega = \mathbf{R}^3$, Ames *et al.* (1989) are unable to establish the above result. They do, however, show $I(t)$ remains bounded below by a constant which is bigger than $I(0)$, as $t \to \infty$.

4.1.5 The Backward in Time Problem for Model III

We recall the basic equations for model III,

$$\frac{\partial v_i}{\partial t} = v_j \frac{\partial v_i}{\partial x_j} + \frac{\partial p}{\partial x_i} - \frac{\partial}{\partial x_j}\left[\left(\nu_0 + \nu_1 |\nabla \mathbf{v}|^2\right)\frac{\partial v_i}{\partial x_j}\right],$$

$$\frac{\partial v_i}{\partial x_i} = 0. \tag{4.1.69}$$

Perhaps surprisingly Ames *et al.* (1989) were able to make least progress with this model. We now investigate this to see why.

Again, it is convenient to rewrite $(4.1.69)_1$ as

$$\frac{\partial \mathbf{v}}{\partial t} = (\mathbf{v} \cdot \nabla)\mathbf{v} + \nabla p + L\mathbf{v} + M\mathbf{v}. \tag{4.1.70}.$$

The operators $L\mathbf{v}$ and $M\mathbf{v}$ in (4.1.70) are defined by

$$L\mathbf{v} = -\nu_0 \Delta \mathbf{v},$$

$$M\mathbf{v} = -\nu_1 \nabla \left(|\nabla \mathbf{v}|^2 \nabla \mathbf{v}\right).$$

At first sight it would appear the method applied to model II should yield exponential growth when $\Omega = \mathbf{R}^2$.

Upon defining $I(t) = \|\mathbf{v}(t)\|^2$, and multiplying (4.1.70) by $\mathbf{v}$ and integration over Ω, we find with use of the boundary conditions,

$$\frac{dI}{dt} = 2\nu_0 \|\nabla \mathbf{v}\|^2 + 2\nu_1 \int_\Omega |\nabla \mathbf{v}|^4 dx. \tag{4.1.71}$$

If we put

$$\Phi = 2\nu_0 \|\nabla \mathbf{v}\|^2 + \nu_1 \int_\Omega |\nabla \mathbf{v}|^4 dx,$$

then

$$\begin{aligned}
\frac{d\Phi}{dt} &= -4\nu_0 (\Delta \mathbf{v}, \mathbf{v}_t) - 4\nu_1 \int_\Omega \nabla\left(|\nabla \mathbf{v}|^2 \nabla \mathbf{v}\right) \cdot \mathbf{v}_t dx, \\
&= -4(\mathbf{v}_t, L\mathbf{v} + M\mathbf{v}), \\
&= 4\|L\mathbf{v} + M\mathbf{v}\|^2 - 4\left(L\mathbf{v} + M\mathbf{v}, \nabla p + (\mathbf{v} \cdot \nabla)\mathbf{v}\right). \tag{4.1.72}
\end{aligned}$$

Again,

$$\left(L\mathbf{v}, \nabla p + (\mathbf{v} \cdot \nabla)\mathbf{v}\right) = 0,$$

since Ω is a two-dimensional region. Further calculation reveals

$$\left(M\mathbf{v}, (\mathbf{v} \cdot \nabla)\mathbf{v}\right) = -\int_{\Omega} \frac{\partial}{\partial x_j}\left(|\nabla\mathbf{v}|^2 \frac{\partial v_i}{\partial x_j}\right) v_k \frac{\partial v_i}{\partial x_k}\, dx,$$

$$= \int_{\Omega} |\nabla\mathbf{v}|^2 \frac{\partial v_i}{\partial x_j}\frac{\partial v_i}{\partial x_k}\frac{\partial v_k}{\partial x_j}\, dx$$

$$+ \int_{\Omega} |\nabla\mathbf{v}|^2 \frac{\partial v_i}{\partial x_j} v_k \frac{\partial^2 v_i}{\partial x_k \partial x_j}\, dx. \qquad (4.1.73)$$

The first term on the right of (4.1.73) is zero since $v_{i,j}v_{i,k}v_{k,j} = 0$. The second term may be rewritten after integration by parts

$$-\frac{1}{2}\int_{\Omega} \frac{\partial}{\partial x_k}\left(|\nabla\mathbf{v}|^2\right) v_k \frac{\partial v_i}{\partial x_j}\frac{\partial v_i}{\partial x_j}\, dx = \frac{1}{4}\int_{\Omega} |\nabla\mathbf{v}|^2 \frac{\partial v_k}{\partial x_k}\, dx = 0.$$

Hence,

$$\left(M\mathbf{v}, (\mathbf{v} \cdot \nabla)\mathbf{v}\right) = 0.$$

But, the trouble arises with the term

$$(M\mathbf{v}, \nabla p).$$

There is no obvious reason why this is zero. Thus, one is henceforth unable to complete the proof as for model II.

One may deduce immediately from (4.1.71) that

$$I(t) \geq I(0).$$

By a non-trivial calculation Ames et al. (1989) are able to deduce that if the solution exists for all t,

$$\lim_{t\to\infty} I(t) \geq \left(\sqrt{I(0)} + \sqrt{\nu_0 \nu_1}\frac{\Phi(0)}{I(0)}\right)^2. \qquad (4.1.74)$$

Also, if $\Omega = \mathbf{R}^3$, Ames et al. (1989) demonstrate

$$\lim_{t\to\infty} \sqrt{I(t)} \geq \sqrt{I(0)} + \frac{2}{3\gamma_1}\left(\frac{\Phi(0)}{I(0)}\right)^{3/4}, \qquad (4.1.75)$$

where γ_1 is a constant depending on ν_0, ν_1, and on a Sobolev embedding constant. However, neither (4.1.74) nor (4.1.75) represent a growth result. The situation for model II is still less than satisfactory.

We close this section by mentioning that Ames et al. (1989) also establish exponential growth results for solutions to the models for a fluid of second or third grade when the domain is the whole space. The models studied are not those dealt with in Sect. 4.2. The theory investigated in the next section involves a generalised second grade fluid model which is capable of describing the phenomena of shear thickening and shear thinning, and has been used to model the flow of glacier ice.

4.2 A Model for a Second Grade Fluid in Glacier Physics

There has been much controversy in the literature over what second and third grade fluid models are, whether they can be exact or whether they are merely approximations to a slow viscoelastic flow. It is not our intention to enter this discussion. We merely point out some mathematical results which demonstrate that when exact models are employed one can prove some non-existence results.

By using arguments of modern continuum thermodynamics, models for a fluid of second grade and of third grade have been derived, respectively, by Dunn & Fosdick (1974) and Fosdick & Rajagopal (1980). These models are comprised of partial differential equations and they consist of the momentum equation,

$$\rho \dot{v}_i = \rho f_i + \frac{\partial}{\partial x_j} T_{ji}, \tag{4.2.1}$$

the equation of continuity,

$$\frac{\partial v_i}{\partial x_i} = 0, \tag{4.2.2}$$

and the rate of work equation,

$$\rho \dot{\epsilon} = T_{ij} L_{ij} - \frac{\partial q_i}{\partial x_i} + \rho r, \tag{4.2.3}$$

where $v_i, \rho, f_i, T_{ij}, r, \epsilon, q_i$ and L_{ij} are, respectively, the velocity, density, body force, stress tensor, heat supply, internal energy, heat flux, and velocity gradient. In these expressions a superposed dot denotes the material derivative. It should be noted that the continuity equation has the reduced form (4.2.2) because the theories of Dunn & Fosdick (1974) and Fosdick & Rajagopal (1980) are for incompressible fluids.

Dunn & Fosdick (1974) adopt a thermodynamic procedure which uses the Clausius–Duhem inequality. They also require that the free energy be a minimum in equilibrium and by this means they are able to deduce that the stress relation for an incompressible, homogeneous fluid of second grade is

$$\mathbf{T} = -p\mathbf{I} + \mu \mathbf{A} + \alpha_1 \mathbf{A}_2 + \alpha_2 \mathbf{A}^2, \tag{4.2.4}$$

where $\mathbf{A}$ and $\mathbf{A}_2$ are the first two Rivlin–Ericksen tensors, defined by

$$\mathbf{A} = \mathbf{L} + \mathbf{L}^T, \qquad \mathbf{A}_2 = \dot{\mathbf{A}} + \mathbf{A}\mathbf{L} + \mathbf{L}^T \mathbf{A}. \tag{4.2.5}$$

By this thermodynamic procedure Dunn & Fosdick (1974) are also able to deduce that the dynamic viscosity μ and the normal stress coefficients α_1 and α_2 appearing in the constitutive law (4.2.4) must satisfy the constraints

$$\mu \geq 0, \qquad \alpha_1 \geq 0, \qquad \alpha_1 + \alpha_2 = 0. \tag{4.2.6}$$

The equivalent thermodynamic procedure for a fluid of third grade was developed by Fosdick & Rajagopal (1980). They show that the constitutive relation for the stress tensor for an incompressible, homogeneous fluid of third grade has form

$$\mathbf{T} = -p\mathbf{I} + \mu\mathbf{A} + \alpha_1\mathbf{A}_2 + \alpha_2\mathbf{A}^2 + \beta(\mathrm{tr}\mathbf{A}^2)\mathbf{A}, \qquad (4.2.7)$$

while the Clausius–Duhem inequality and the requirement that the free energy be a minimum in equilibrium imposes the following constraints on the dynamic viscosity μ, the normal stress coefficients α_1 and α_2, and the coefficient β,

$$\mu \geq 0, \qquad \alpha_1 \geq 0, \qquad \beta \geq 0, \qquad |\alpha_1 + \alpha_2| \leq \sqrt{24\mu\beta}. \qquad (4.2.8)$$

Man & Sun (1987) and Man (1992) use a second grade fluid theory in an analysis of glacier flow. To encompass real effects observed in glaciological field studies they require a model capable of producing the effects of shear thickening and shear thinning and so they must generalise the stress law (4.2.4). To this end they propose that the constitutive equation (4.2.4) be replaced by

$$\mathbf{T} = -p\mathbf{I} + \mu\Pi^{m/2}\mathbf{A} + \alpha_1\mathbf{A}_2 + \alpha_2\mathbf{A}^2, \qquad (4.2.9)$$

where $\Pi = \frac{1}{2}\mathrm{tr}\mathbf{A}^2$, and m is a real number. The idea is that equation (4.2.9) is a combination of the classical power law viscoelastic fluid together with that for a fluid of second grade. Shear thickening means that the viscosity increases with increasing shear velocity, and vice-versa, shear thinning occurs when the viscosity decreases as the shear velocity is increased. The model (4.2.9) incorporates both of these effects in that if $\mu\Pi^{m/2}$ is interpreted as viscosity then (4.2.9) predicts shear thickening for $m > 0$ and shear thinning for $m < 0$. The work of Man (1992) is devoted to the existence and uniqueness question for a weak solution to the boundary initial value problem for model (4.2.9). Man (1992) requires that the dynamic viscosity, normal stress coefficient α_1, and shear thickening/thinning number m be such that $\alpha_1 > 0$, $\mu > 0$, and $-1 < m < 0$. The paper of Man & Sun (1987) considers another generalized second grade fluid which has stress relation

$$\mathbf{T} = -p\mathbf{I} + \Pi^{m/2}(\mu\mathbf{A} + \alpha_1\mathbf{A}_2 + \alpha_2\mathbf{A}^2). \qquad (4.2.10)$$

We shall here refer to the models consisting of (4.2.1) - (4.2.3) together with the constitutive laws (4.2.9) or (4.2.10) as model I or model II, respectively.

Gupta & Massoudi (1993) employ the models of Man & Sun (1987), although they make the realistic assumption that the dynamic viscosity μ is a decreasing exponential function of temperature, T. This model is also employed to study thermal convection by Franchi & Straughan (1993). In this section we decribe work of Franchi & Straughan (1993) who are interested in three-space-dimension problems, although we only include an exposition of their work on non-existence and not on thermal convection. Franchi &

Straughan only develop a blow-up result when the constitutive law (4.2.9) holds. However, we here include a growth result also for the case of (4.2.10), in the event that the normal stress coefficient $\alpha_1 < 0$ and the problem forward in time is under investigation. When we study the backward in time problem for model II we derive an upper bound showing that a solution is suitably bounded above by a growing exponential, thereby preventing the possibility of finite time blow-up. Henceforth, we assume that the equations are rescaled so that the density has value one; this causes no loss of generality.

The focus of this article is on non-existence results. In this regard Fosdick & Straughan (1981) showed that for the third grade fluid model (4.2.7), the condition $\alpha_1 < 0$ ensures that solutions to the boundary initial value problem with zero boundary data cannot exist globally. Straughan (1987) demonstrates that the condition $\alpha_1 > 0$ ensures non-existence for the boundary initial value problem when time is reversed, i.e. the backward in time problem. The non-existence results of Fosdick & Straughan (1981) and Straughan (1987) hinge on the presence of the non-linearity associated with the β term which is present in the third grade model (4.2.7) but not in the standard second grade model represented by (4.2.4).

4.2.1 Non-existence Forward in Time for Model I

The paper of Franchi & Straughan (1993) proves global non-existence forward in time for the generalised second grade fluid model I with constitutive law (4.2.9), *when* $\alpha_1 < 0$. They study the boundary initial value problem for (4.2.1) and (4.2.2), together with the constitutive law (4.2.9) for $\mu, \alpha_1, \alpha_2, m$ constant, with $\mu > 0$, $\alpha_1 < 0$ and $m > 0$. The spatial domain Ω is a bounded region in three-space and on the boundary of Ω, Γ, it is assumed that

$$v_i = 0, \qquad \mathbf{x} \in \Gamma. \tag{4.2.11}$$

The initial data is given by

$$\mathbf{v}(\mathbf{x}, 0) = \mathbf{v}_0(\mathbf{x}). \tag{4.2.12}$$

The situation analysed by Franchi & Straughan (1993) corresponds to isolated cannister flow in which a container of generalised second fluid is subjected to an arbitrary motion and then held fixed at time $t = 0$ and thereafter.

The proof of non-existence in Franchi & Straughan (1993) develops a differential inequality for the functional $H(t)$ given by

$$H(t) = \frac{1}{2}\zeta\|\mathbf{A}(t)\|^2 - \|\mathbf{v}(t)\|^2, \tag{4.2.13}$$

where we have set $\zeta = -\alpha_1 (> 0)$.

The proof begins by forming an energy equation from (4.2.1) by multiplying by v_i and integrating over Ω. With the aid of the boundary condition (4.2.11) and the constitutive equation (4.2.9), we see that

$$\frac{d}{dt}\left(\|\mathbf{v}\|^2 + \frac{1}{2}\alpha_1\|\mathbf{A}\|^2\right) = -2\mu\int_\Omega \Pi^{m/2}A_{ij}v_{i,j}\,dx$$

$$-(\alpha_1 + \alpha_2)\int_\Omega \mathrm{tr}\mathbf{A}^3\,dx. \qquad (4.2.14)$$

It is assumed that the normal stress coefficient α_1 satisfies the condition $\alpha_1 < 0$, and both normal stress coefficients α_1 and α_2 satisfy the relation $(1.6)_2$ of Dunn & Fosdick (1974), whereby $\alpha_1 + \alpha_2 = 0$. Thus, from (4.2.14) we deduce

$$\frac{d}{dt}\left(\|\mathbf{v}\|^2 + \frac{1}{2}\alpha_1\|\mathbf{A}\|^2\right) = -2\mu\int_\Omega \Pi^{m/2}A_{ij}v_{i,j}\,dx. \qquad (4.2.15)$$

Due to the fact that the tensor $\mathbf{A}$ is symmetric, we write

$$\Pi^{m/2}A_{ij}v_{i,j} = \frac{1}{2}\Pi^{m/2}\mathrm{tr}\mathbf{A}^2,$$

and so since $\Pi = \frac{1}{2}\mathrm{tr}\mathbf{A}^2$, from (4.2.15) it follows that

$$\frac{d}{dt}\left(\|\mathbf{v}\|^2 + \frac{1}{2}\alpha_1\|\mathbf{A}\|^2\right) = -2\mu\int_\Omega \Pi^{1+m/2}\,dx. \qquad (4.2.16)$$

To proceed from this the last term is bounded below using Hölder's inequality by the relation

$$\int_\Omega \Pi^{1+m/2}\,dx \geq c\left(\int_\Omega \Pi\,dx\right)^{\frac{2+m}{2}}, \qquad (4.2.17)$$

in which the constant c is given by

$$c = \frac{1}{[M(\Omega)]^{(2+m)^2/2m}},$$

where $M(\Omega)$ is the Lebesgue measure of Ω.

We thus multiply (4.2.16) by -1, set $\zeta = -\alpha_1(> 0)$ and use inequality (4.2.17) to derive

$$\frac{d}{dt}\left(\frac{1}{2}\zeta\|\mathbf{A}\|^2 - \|\mathbf{v}\|^2\right) = \mu\int_\Omega \Pi^{1+m/2}\,dx,$$

$$\geq \mu c 2^{-m/2}\|\mathbf{A}\|^{2(1+m/2)}. \qquad (4.2.18)$$

For the class of non-existence of solutions we shall require that the initial data satisfy

$$\frac{1}{2}\zeta\|\mathbf{A}(0)\|^2 - \|\mathbf{v}(0)\|^2 > 0. \qquad (4.2.19)$$

From inequality (4.2.18) the function $H(t)$ given by (4.2.13) is increasing and since $H(0) > 0$ it follows that

$$\frac{1}{2}\zeta\|\mathbf{A}(t)\|^2 - \|\mathbf{v}(t)\|^2 > 0,$$

for all $t > 0$ for which the solution exists. Therefore,

$$\frac{1}{2}\zeta\|\mathbf{A}\|^2 \geq \frac{1}{2}\zeta\|\mathbf{A}\|^2 - \|\mathbf{v}\|^2 > 0,$$

and so from inequality (4.2.18) we may arrive at

$$\frac{d}{dt}\left(\frac{1}{2}\zeta\|\mathbf{A}\|^2 - \|\mathbf{v}\|^2\right) \geq k\left(\frac{1}{2}\zeta\|\mathbf{A}\|^2 - \|\mathbf{v}\|^2\right)^{1+m/2}, \qquad (4.2.20)$$

where the constant k is given by

$$k = \frac{\mu 2^{2+m/2}}{|\alpha_1|^{(2+m)} M^{(2+m)^2/2m}}. \qquad (4.2.21)$$

We integrate inequality (4.2.20) to derive

$$H(t) \geq \frac{2^{2/m} H(0)}{\left(2 - km(H(0))^{m/2} t\right)^{2/m}}, \qquad (4.2.22)$$

where the function H is defined in (4.2.13). Expression (4.2.22) is a lower bound for the solution to the generalised second grade problem with constitutive eqiation (4.2.9) when $\alpha_1 < 0$, in the measure $H(t)$. It shows that the solution to (4.2.1), (4.2.2), and (4.2.9) cannot exist globally in time. From the right hand side of (4.2.22) an upper bound for the length of the possible existence interval is T^* which is given by

$$T^* = \frac{2}{km\left(\frac{1}{2}\zeta\|\mathbf{A}(0)\|^2 - \|\mathbf{v}(0)\|^2\right)^{m/2}}. \qquad (4.2.23)$$

4.2.2 Non-existence Backward in Time for Model I

In this subsection we consider that part of the work of Franchi & Straughan (1993) where they prove global non-existence backward in time for the generalised second grade fluid model I under the restrictoin $\alpha_1 > 0$. Thus, we investigate the cannister flow problem for a generalised second grade fluid with constitutive equation (4.2.9), but backward in time. The conditions imposed on the coefficients are that

$$\alpha_1 > 0, \qquad m > 0, \qquad \alpha_1 + \alpha_2 = 0. \qquad (4.2.24)$$

The body force is again supposed derivable from a potential and, therefore, absorbed in the pressure. The backward in time equations satisfied by the generalised second grade fluid in isothermal flow are

$$\rho\left(\frac{\partial v_i}{\partial t} - v_j\frac{\partial v_i}{\partial x_j}\right) = -\frac{\partial}{\partial x_j}T_{ji}, \tag{4.2.25}$$

$$\frac{\partial v_i}{\partial x_i} = 0, \tag{4.2.26}$$

where T_{ji} is given by (4.2.9), although it is to be understood that time is reversed in the Rivlin–Ericksen tensor $\mathbf{A}_2$.

We commence by multiplying (4.2.25) by v_i and integrate over Ω. If we do this and then use inequality (4.2.17) we may obtain

$$\frac{d}{dt}\left(\|\mathbf{v}\|^2 + \frac{1}{2}\alpha_1\|\mathbf{A}\|^2\right) = \mu\int_\Omega \Pi^{1+m/2}\,dx,$$

$$\geq 2c\mu\left(\int_\Omega \Pi\,dx\right)^{1+m/2}. \tag{4.2.27}$$

The domain Ω is bounded and since $v_i = 0$ on Γ, Poincaré's inequality, for a positive constant λ_1, shows that

$$\int_\Omega \Pi\,dx = \frac{1}{2}\int_\Omega \mathrm{tr}\mathbf{A}^2\,dx = \frac{1}{2}\|\mathbf{A}\|^2 \geq \frac{1}{2}\lambda_1\|\mathbf{v}\|^2. \tag{4.2.28}$$

By use of this inequality we may now show

$$\int_\Omega \Pi\,dx \geq \left(\frac{\lambda_1}{2+\alpha_1\lambda_1}\right)\left(\|\mathbf{v}\|^2 + \frac{1}{2}\alpha_1\|\mathbf{A}\|^2\right), \tag{4.2.29}$$

and hence from (4.2.27),

$$\frac{d}{dt}\left(\|\mathbf{v}\|^2 + \frac{1}{2}\alpha_1\|\mathbf{A}\|^2\right)$$

$$\geq 2c\mu\left(\frac{\lambda_1}{2+\alpha_1\lambda_1}\right)^{1+m/2}\left(\|\mathbf{v}\|^2 + \frac{1}{2}\alpha_1\|\mathbf{A}\|^2\right)^{1+m/2}. \tag{4.2.30}$$

If we now define the function $I(t)$ by

$$I(t) = \|\mathbf{v}(t)\|^2 + \frac{1}{2}\alpha_1\|\mathbf{A}(t)\|^2,$$

then inequality (4.2.30) may be alternatively written as

$$\frac{dI}{dt} \geq k_1 I^{1+m/2}, \tag{4.2.31}$$

where the constant k_1 is given by

$$k_1 = \frac{2\mu}{M^{(2+m)^2/2m}}\left(\frac{\lambda_1}{2+\alpha_1\lambda_1}\right)^{1+m/2}.$$

Inequality (4.2.31) is integrated to deduce

$$I(t) \geq \frac{2^{2/m}I(0)}{\left(2 - k_1 m (I(0))^{m/2} t\right)^{2/m}}.\tag{4.2.32}$$

The inequality (4.2.32) represents a lower bound for the solution to the generalised second grade fluid cannister flow problem with $\alpha_1 > 0$ and constitutive law (4.2.9). Global non-existence follows from (4.2.32) and a bound for the maximal interval of existence is

$$T^{**} = \frac{2}{k_1 m \left(\frac{1}{2}\alpha_1 \|\mathbf{A}(0)\|^2 + \|\mathbf{v}(0)\|^2\right)^{m/2}}.$$

4.2.3 Exponential Growth Forward in Time for Model II

The equations for a generalised second grade fluid with the constitutive law (4.2.10) in isothermal conditions may be derived from the partial differential equations (4.2.1) and (4.2.2), and are for a body force f_i given by a potential,

$$\rho\left(\frac{\partial v_i}{\partial t} + v_j \frac{\partial v_i}{\partial x_j}\right) = -\frac{\partial p}{\partial x_i} + \mu \frac{\partial}{\partial x_j}\left(\Pi^{m/2} A_{ji}\right)$$

$$+ \alpha_1 \frac{\partial}{\partial x_j}\left[\left(\frac{\partial A_{ij}}{\partial t} + v_k \frac{\partial A_{ij}}{\partial x_k}\right)\Pi^{m/2}\right]$$

$$+ \alpha_1 \frac{\partial}{\partial x_j}\left[\left(A_{ir} v_{r,j} + v_{r,i} A_{rj}\right)\Pi^{m/2}\right]$$

$$+ \alpha_2 \frac{\partial}{\partial x_j}\left(A_{is} A_{sj} \Pi^{m/2}\right),\tag{4.2.33}$$

$$\frac{\partial v_i}{\partial x_i} = 0,\tag{4.2.34}$$

where we have used the form (4.2.5) for the Rivlin–Ericksen tensor $\mathbf{A}_2$, and we have explicitly included the density ρ even though we shall treat it subsequently as having the numerical value of one.

We adopt the Dunn & Fosdick (1974) condition

$$\alpha_1 + \alpha_2 = 0,\tag{4.2.35}$$

and in this subsection we assume

$$\alpha_1 < 0.\tag{4.2.36}$$

We shall also be considering the case $m > 0$. Equations (4.2.33), (4.2.34) are defined on the region $\Omega \times (0, \infty)$ for Ω a bounded domain, and the boundary conditions are again those appropriate for isolated cannister flow, i.e.

$$v_i = 0, \qquad \text{on } \Gamma.\tag{4.2.37}$$

The solution v_i is subject to initial data

$$v_i(\mathbf{x}, 0) = v_{i0}(\mathbf{x}). \tag{4.2.38}$$

The starting point is to multiply (4.2.33) by v_i, integrate over Ω, use (4.2.37) and (4.2.34), to find

$$\frac{d}{dt}\frac{1}{2}\|\mathbf{v}\|^2 = -\mu \int_\Omega \Pi^{m/2} A_{ij} v_{i,j}\, dx - \alpha_1 \int_\Omega \Pi^{m/2}\frac{\partial A_{ij}}{\partial t} v_{i,j}\, dx$$

$$- \alpha_1 \int_\Omega \Pi^{m/2} v_k \frac{\partial A_{ij}}{\partial x_k} v_{i,j}\, dx$$

$$- \alpha_1 \int_\Omega \Pi^{m/2}\left(A_{ir} v_{r,j} v_{i,j} + A_{rj} v_{r,i} v_{i,j}\right) dx$$

$$- \alpha_2 \int_\Omega \Pi^{m/2} A_{is} A_{sj} v_{i,j}\, dx. \tag{4.2.39}$$

In the next stages we repeatedly use the facts that $A_{ij} = v_{i,j} + v_{j,i}$, that A_{ij} is symmetric, that $\Pi = \frac{1}{2}\mathrm{tr}\mathbf{A}^2$, and that $A_{ij} v_{i,j} = \Pi$. We may show

$$\Pi^{m/2} A_{ij} v_{i,j} = \Pi^{1+m/2}, \tag{4.2.40}$$

$$\int_\Omega \Pi^{m/2}\frac{\partial A_{ij}}{\partial t} v_{i,j}\, dx = \frac{1}{2}\int_\Omega \Pi^{m/2}\frac{\partial A_{ij}}{\partial t} A_{ij}\, dx,$$

$$= \frac{1}{2}\int_\Omega \Pi^{m/2}\frac{\partial \Pi}{\partial t}\, dx,$$

$$= \frac{1}{(m+2)}\frac{d}{dt}\int_\Omega \Pi^{1+m/2}\, dx, \tag{4.2.41}$$

where in the last line we have used the fact that $v_i = 0$ on Γ.

Moreover,

$$\int_\Omega \Pi^{m/2} v_k \frac{\partial A_{ij}}{\partial x_k} v_{i,j}\, dx = \frac{1}{2}\int_\Omega \Pi^{m/2} v_k \frac{\partial}{\partial x_k} A_{ij} A_{ij}\, dx,$$

$$= \int_\Omega \Pi^{m/2} v_k \frac{\partial \Pi}{\partial x_k}\, dx,$$

$$= \frac{2}{(m+2)}\int_\Omega v_k \frac{\partial}{\partial x_k}\Pi^{1+m/2}\, dx,$$

$$= 0, \tag{4.2.42}$$

since $v_i = 0$ on Γ and (4.2.34) holds.

Finally, by symmetry in i and j,

$$A_{is} A_{sj} v_{i,j} = \frac{1}{2} A_{is} A_{sj} A_{ij}, \tag{4.2.43}$$

and also by symmetry in the appropriate indices,

$$
\begin{aligned}
A_{ir}v_{r,j}v_{i,j} + A_{rj}v_{r,i}v_{i,j} &= \frac{1}{2}\left(A_{ir}v_{r,j} + A_{rj}v_{r,i}\right)A_{ij}, \\
&= \frac{1}{2}\left(A_{ir}A_{ij}v_{r,j} + A_{rj}A_{ij}v_{r,i}\right), \\
&= \frac{1}{2}\left(A_{ir}A_{rj}A_{ij}\right).
\end{aligned}
\tag{4.2.44}
$$

We now combine (4.2.40) - (4.2.44) together in (4.2.39) to see that

$$
\frac{d}{dt}\left(\frac{1}{2}\|\mathbf{v}\|^2 + \frac{\alpha_1}{(m+2)}\int_\Omega \Pi^{1+m/2}\,dx\right) = -\mu\int_\Omega \Pi^{1+m/2}\,dx
$$
$$
- \frac{1}{2}(\alpha_1 + \alpha_2)\int_\Omega \Pi^{m/2}A_{is}A_{sj}A_{ij}\,dx.
\tag{4.2.45}
$$

The last term is zero thanks to (4.2.35). Now put $\zeta = -\alpha_1 (> 0)$ and then from (4.2.45) we have

$$
\frac{d}{dt}\left(\frac{\zeta}{(m+2)}\int_\Omega \Pi^{1+m/2}\,dx - \frac{1}{2}\|\mathbf{v}\|^2\right) = \mu\int_\Omega \Pi^{1+m/2}\,dx.
\tag{4.2.46}
$$

If we now define the function $P(t)$ by

$$
P(t) = \frac{\zeta}{(m+2)}\int_\Omega \Pi^{1+m/2}\,dx - \frac{1}{2}\|\mathbf{v}\|^2,
$$

then from (4.2.46) it follows that

$$
\frac{dP}{dt} \geq \left[\frac{\mu(m+2)}{\zeta}\right]P.
\tag{4.2.47}
$$

By integration one then derives the lower bound

$$
P(t) \geq P(0)\exp\left\{\frac{\mu(m+2)t}{\zeta}\right\}.
\tag{4.2.48}
$$

Provided the initial data are such that

$$
P(0) = \frac{-\alpha_1}{(m+2)2^{(2+m)/2}}\int_\Omega \left[(v_{i,j}^0 + v_{j,i}^0)(v_{i,j}^0 + v_{j,i}^0)\right]dx
$$
$$
- \frac{1}{2}\|\mathbf{v}_0\|^2 > 0,
\tag{4.2.49}
$$

the solution to the isolated cannister flow problem for model II must grow at least exponentially.

4.2.4 Exponential Boundedness Backward in Time for Model II

The situation here is as in the previous subsection except we suppose

$$\alpha_1 > 0, \tag{4.2.50}$$

and time is reversed in (4.2.33). Thus, we shall again consider isolated cannister flow for the generalised second grade fluid model II, i.e. with the constitutive law (4.2.10).

The steps leading to (4.2.46) are repeated, *mutatis mutandis*. One may arrive at the equation,

$$\frac{d}{dt}\left(\frac{1}{2}\|\mathbf{v}\|^2 + \frac{\alpha_1}{(m+2)}\int_\Omega \Pi^{1+m/2}\,dx\right) = \mu\int_\Omega \Pi^{1+m/2}\,dx. \tag{4.2.51}$$

Rather than try to establish a growth estimate for a solution from equation (4.2.51) it is now straightforward to derive an upper bound. Upon setting

$$Q(t) = \frac{1}{2}\|\mathbf{v}\|^2 + \frac{\alpha_1}{(m+2)}\int_\Omega \Pi^{1+m/2}\,dx,$$

one easily sees from (4.2.51) that

$$\frac{dQ}{dt} \le \frac{\mu(m+2)}{\alpha_1}Q. \tag{4.2.52}$$

Therefore, by integrating (4.2.52) we find

$$Q(t) \le Q(0)\exp\left[\frac{\mu(m+2)t}{\alpha_1}\right]. \tag{4.2.53}$$

While (4.2.53) says nothing about growth of solutions it is useful in that it shows the solution to the isolated cannister flow problem for model II backward in time cannot blow-up in finite time, at least not in the measure defined by $Q(t)$.

4.3 Blow-Up for Generalised KdeV Equations

Bona *et al.* (1992, 1995) and Bona & Saut (1993) treat the generalised Korteweg–de Vries (GKdV) equation

$$\frac{\partial u}{\partial t} + u^p\frac{\partial u}{\partial x} + \epsilon\frac{\partial^3 u}{\partial x^3} = 0, \tag{4.3.1}$$

for p a positive integer, and Bona *et al.* (1992) develop a generalisation of (4.3.1), namely the generalised Korteweg–de Vries–Burgers (GKdVB) equation,

$$\frac{\partial u}{\partial t} + u^p \frac{\partial u}{\partial x} - \delta \frac{\partial^2 u}{\partial x^2} + \epsilon \frac{\partial^3 u}{\partial x^3} = 0. \qquad (4.3.2)$$

Equations (4.3.1), (4.3.2) hold for $t > 0$ on the whole real line $x \in \mathbf{R}$ and are subject to initial data

$$u(x,0) = u_0(x), \qquad (4.3.3)$$

which in Bona et al. (1992, 1995) is primarily a periodic function of the variable x. Many other pertinent references to studies of (4.3.1) and (4.3.2) are given in Bona et al. (1992, 1995) and in Bona & Saut (1993).

Equations (4.3.1), (4.3.2) have a very wide application in mechanics with $p = 1, \delta = 0$ being the famous KdV equation well known in water waves and several other areas of mechanics. As pointed out by Bona et al. (1992), when $p = 1, \delta, \epsilon > 0$, the equation (4.3.2) has also wide application as has the case $p = 2$. The famous paper of Benjamin et al. (1972) considers physical applications also when $p \geq 2$.

In Bona et al. (1992, 1995) the emphasis is on developing accurate numerical schemes to study the evolution of a solution to (4.3.1) or (4.3.2), and in particular, to follow the solution in time as it approaches blow-up. Of course, to study blow-up of a solution to a partial differential equation numerically is a very difficult thing to do and Bona et al. (1992, 1995) are careful to incorporate analytical checks as they proceed. Their numerical techniques are based on a smooth periodic spline Galerkin finite element spatial discretization together with a Runge–Kutta implicit method in time. Due to the need to follow rapid solution growth they develop an adaptive code which takes into account the local refinement of the spatial grid, changing time step size, and also spatial translations of the solution. To control the numerical solution Bona et al. (1992,1995) note that the functional

$$I_3(v) = \int_0^1 \left[v^{p+2} - \frac{1}{2}\epsilon(p+1)(p+2)\left(\frac{\partial v}{\partial x}\right)^2 \right] dx, \qquad (4.3.4)$$

for period 1, is time invariant for equation (4.3.1). They make substantial use of this invariant as a tolerance in their adaptive numerical scheme.

Bona et al. (1995) is concerned with (4.3.1) whereas Bona et al. (1992) is largely devoted to a treatment of (4.3.2). According to Bona et al. (1992) the initial value problem for (4.3.1) is globally well posed for $p < 4$ but for $p \geq 4$ solutions may possibly blow-up in finite time. The case of $p = 5$ is investigated in detail and computations are started with the initial data function being a slightly perturbed solitary wave

$$u_0(x) = \lambda A \operatorname{sech}^{2/p}\left[K\left(x - \frac{1}{2}\right)\right],$$

for suitable parameters λ, A, K. The computations of Bona et al. (1992, 1995) suggest the solution to (4.3.1) behaves like

$$u(x, t) = \frac{1}{(T-t)^{2/3p}} \, \phi\left(\frac{X-x}{(T-t)^{1/3}}\right) + \text{bounded term}, \qquad (4.3.5)$$

for ϕ a bounded function. The blow-up rate is thus $2/3p$ and this is computed in various L^q norms for u and u_x. The blow-up rate of the numerical solution is compared with that of (4.3.5) and excellent agreement is seen. When $\delta \neq 0$, but small, i.e. (4.3.2) is considered, with $p = 5$ and other parameters the same, blow-up would again appear evident. Thus, the effect of dissipation via the δ term does not prevent blow-up. Indeed, the blow-up rates for $\delta = 0$ and $\delta \neq 0$ appear the same, although when $\delta \neq 0$ the time of blow-up is delayed. This is to be expected since dissipation should have an inhibiting effect. Bona *et al.* (1992) also demonstrate that when δ is large blow-up will not occur and the solution even decays. In fact, they show that the threshold parameter which governs whether solutions blow-up or decay is $C = \delta^2/\epsilon A^p$. For C^* a threshold solutions blow-up in finite time when $C < C^*$ and are bounded and persist for all time for $C > C^*$. This part of the work is based on careful computation, although an analytical result guaranteeing decay is given.

Bona & Saut (1993) presents a rigorous analysis for equation (4.3.1) subject to an arbitrary initial data function

$$u_0(x) = \psi(x).$$

However, the main thrust of Bona & Saut (1993) is to investigate another kind of blow-up, a phenomenon they entitle dispersive blow-up. They refer to the blow-up for $p \geq 4$ as non-linear blow-up since it is due to the non-linear term and in that case the L^2 or other L^p norms typically blow-up in finite time. Roughly speaking, for dispersive blow-up the solution becomes singular at a single point (x^*, t^*) in such a way that the L^p norms remain finite.

The motivation of the work of Bona & Saut (1993) would appear to be to rigorously establish for the non-linear equation (4.3.1) a supposition of Benjamin *et al.* (1972). This observes that for $p = 0$ (the linear equation) in (4.3.1), a solution like $\cos(kx - \omega t)$ leads to the dispersion relation

$$\omega(k) = k(1 - k^2). \qquad (4.3.6)$$

Thus the frequency ω, and the group velocity $c_g = \omega'(k)$ together with the phase velocity $c = \omega(k)/k$ are unbounded. Hence, many short waves may coalesce at a single point to result in a loss of solution smoothness, i.e. a focussing type of blow-up. The work of Bona & Saut (1993) thus addresses this type of blow-up which can occur for $p \geq 0$ and is associated with the linearised dispersion equation for (4.3.1). This type of blow-up is called dispersive blow-up by Bona & Saut (1993).

The paper of Bona & Saut (1993) contains an authoritative analysis and shows that for $k \geq 2$ with initial data $\psi \in H^k(\mathbf{R})$, $p < 4$ guarantees a global solution which stays spatially in H^k whereas for $p \geq 4$ restrictions must be placed on a suitable norm of the initial data. An exact statement is given in

Theorem 2.3 of Bona & Saut (1993). Interestingly, Bona & Saut (1993) also show that the problem

$$\frac{\partial u}{\partial t} - u^{2q}\frac{\partial u}{\partial x} + \frac{\partial^3 u}{\partial x^3} = 0, \qquad t > 0, x \in \mathbf{R},$$

$$u(x,0) = \psi(x), \tag{4.3.7}$$

for q any positive integer, has a global solution.

Bona & Saut (1993) establish several existence results for solutions in weighted Sobolev spaces. These results have a major bearing on their dispersive blow-up theory. The dispersive blow-up results of Bona & Saut (1993) are proved by a very interesting method and involve delicate calculations and estimates with Airy functions,

$$Ai(x) = \frac{1}{\pi}\int_0^\infty \cos\left(\frac{1}{3}\theta^3 + \theta x\right) d\theta.$$

Basically, however, they show that a solution to (4.3.1) can exist with u in $H^k(\mathbf{R})$ in x, $\partial^k u/\partial x^k$ continuous on $\mathbf{R} \times [0,T] - \{(x^*, t^*)\}$ but

$$\lim_{x\to x^*, t\to t^*} \frac{\partial^k u}{\partial x^k}(x,t) = +\infty.$$

Bona & Saut (1993) extend their results to systems generalising (4.3.1) which have a more general dispersion relation of form,

$$\omega(k) = kP(k), \tag{4.3.8}$$

where P is an even degree polynomial in k. They note that this is physically relevant since $P(k) = 1 - k^2 + k^4$ is appropriate to the next order approximation for the linearised dispersion relation for the equations for surface waves on water which lead to the KdV equation. To illustrate the extension of their dispersive blow-up theory to equations which will yield (4.3.8), Bona & Saut (1993) concentrate on the partial differential equation

$$\frac{\partial u}{\partial t} + u^p\frac{\partial u}{\partial x} + \frac{\partial^5 u}{\partial x^5} = 0. \tag{4.3.9}$$

This equation has direct application for small amplitude, long waves on a shallow layer of water, as observed by Bona & Saut (1993). Bona & Saut (1993) show that $p < 8$ yields a globally well posed problem for the partial differential equation (4.3.9). However, they show dispersive blow-up can occur when $p = 1$.

4.4 Very Rapid Growth in Ferrohydrodynamics

Malik & Singh (1992) study the development of a surface instability between two ferromagnetic fluids. The physical situation they investigate involves two inviscid, incompressible, ferromagnetic fluids of densities ρ_1 and ρ_2 which occupy the half spaces $z < 0$ and $z > 0$, respectively; these can be thought of as regions 1 and 2. Gravity acts in the negative z-direction and a constant magnetic field, $\mathbf{H}$, acts in the upward vertical direction. Since both fluids are ferromagnetic the magnetic induction field $\mathbf{B}$ satisfies the non-linear law $\mathbf{B} = \mu(\mathbf{H})\mathbf{H}$ in each fluid, which has its own magnetic permeability μ_1 or μ_2, according to whether we are in region 1 or 2. Malik & Singh (1992) restrict attention to two-dimensional disturbances and suppose the velocities remain irrotational. Since for a ferrofluid the magnetostatic limit is assumed, cf. Rosensweig (1985) p. 91, Straughan (1992) p. 177, then the magnetic field satisfies the partial differential equation

$$\nabla \times \mathbf{H} = \mathbf{0},$$

and then a magnetic potential ψ exists such that $\mathbf{H} = -\nabla\psi$. The irrotational flow field allows the introduction of a velocity potential ϕ such that $\mathbf{v} = \nabla\phi$. Allowing for interface motion with $\eta(x,t)$ being the surface change from $z = 0$, the equations governing the velocity and magnetic fields in the semi-infinite regions become

$$\Delta\phi^{(1)} = 0, \qquad \Delta\psi^{(1)} = 0, \qquad -\infty < z < \eta(x,t), \qquad (4.4.1)$$
$$\Delta\phi^{(2)} = 0, \qquad \Delta\psi^{(2)} = 0, \qquad \eta(x,t) < z < \infty. \qquad (4.4.2)$$

Malik & Singh (1992) require $\phi^{(i)}$ and $\psi^{(i)}$ to satisfy the boundary conditions at infinity,

$$|\nabla\phi^{(1)}| \to 0, \qquad |\nabla\psi^{(1)}| \to 0, \qquad z \to -\infty,$$
$$|\nabla\phi^{(2)}| \to 0, \qquad |\nabla\psi^{(2)}| \to 0, \qquad z \to \infty. \qquad (4.4.3)$$

The interfacial boundary conditions Malik & Singh (1992) employ consist of a kinematic condition

$$\frac{\partial\eta}{\partial t} + \mathbf{v}^{(m)} \cdot \nabla\eta = w^{(m)}, \qquad m = 1, 2, \qquad (4.4.4)$$

where $w^{(m)} = v_3^{(m)}$, continuity of the normal component of the magnetic induction,

$$(\mathbf{B}^{(1)} - \mathbf{B}^{(2)}) \cdot \mathbf{n} = 0, \qquad (4.4.5)$$

$\mathbf{n}$ being the unit normal to the interface, and continuity of the tangential component of $\mathbf{H}$, i.e.

$$(\mathbf{H}^{(1)} - \mathbf{H}^{(2)}) \times \mathbf{n} = \mathbf{0}. \qquad (4.4.6)$$

In addition, the jump in stress at the interface is balanced by surface tension. In general, this condition may be written as

$$t^i = t_{ji}n^j = \sigma b_\alpha^\alpha n^i + a_{\alpha\beta}x^i_{;\alpha}\sigma_{;\beta}\,, \tag{4.4.7}$$

where $t^i, t_{ji}, \sigma, b_\alpha^\alpha, a_{\alpha\beta}$ are the stress vector, stress tensor, surface tension, mean curvature, and first fundamental form of the surface.

Malik & Singh (1992) consider a constant surface tension and thus reduce equation (4.4.7) to the form

$$\left[\rho_1\frac{\partial\phi^{(1)}}{\partial t} - \rho_2\frac{\partial\phi^{(2)}}{\partial t} + (\rho_1 - \rho_2)g\eta + \frac{1}{2}\left(\rho_1|\nabla\phi^{(1)}|^2 - \rho_2|\nabla\phi^{(2)}|^2\right)\right]n_i$$
$$+ (m_{ij}^{(1)} - m_{ij}^{(2)})n_j = \sigma\,\frac{\eta_{xx}}{(1+\eta_x^2)^{3/2}}\,. \tag{4.4.8}$$

The last term represents the curvature contribution and m_{ij} is the magnetic stress tensor in each fluid which has form

$$m_{ij} = -\frac{1}{4\pi}\int_0^H \left[\mu - \rho\left(\frac{\partial\mu}{\partial\rho}\right)_H\right]H_j\,dH_i + \frac{\mu}{4\pi}H_iH_j\,. \tag{4.4.9}$$

As Malik & Singh (1992) show, (4.4.4) and (4.4.5) become

$$\frac{\partial\eta}{\partial t} + \frac{\partial\phi^{(k)}}{\partial x}\frac{\partial\eta}{\partial x} = \frac{\partial\phi^{(k)}}{\partial z}, \qquad k = 1,2, \tag{4.4.10}$$

and

$$\mu_1\mathbf{H}^{(1)}\cdot\mathbf{n} = \mu_2\mathbf{H}^{(2)}\cdot\mathbf{n}. \tag{4.4.11}$$

To develop an instability analysis about a bifurcation value of $|\mathbf{H}|$, H_m, Malik & Singh (1992) perform a formal expansion in terms of a small parameter

$$\epsilon = \sqrt{\frac{H^2}{H_m^2} - 1}\,.$$

Their method uses a multiple scales expansion in the length and time scales $x_n = \epsilon^n x, t_n = \epsilon^n t$. The potentials and interface variables are expanded as

$$\phi(x, z, t) = \sum_{n=1}^{3} \epsilon^{n+1}\phi_{n+1} + O(\epsilon^5),$$

$$\psi(x, z, t) = \sum_{n=1}^{3} \epsilon^{n+1}\psi_{n+1} + O(\epsilon^5), \tag{4.4.12}$$

$$\eta(x, z, t) = \sum_{n=1}^{3} \epsilon^{n+1}\eta_{n+1} + O(\epsilon^5),$$

where ϕ_n, ψ_n, η_n depend on lower scales. Expressions (4.4.12) are substituted into (4.4.1)–(4.4.11) and a system of equations is obtained by equating like powers of ϵ. We do not give all the technical details and refer to Malik & Singh (1992) for this. However, they study the $O(\epsilon^2), O(\epsilon^3)$ and $O(\epsilon^4)$ problems which are sufficient for their purpose.

By neglecting the second harmonic resonance Malik & Singh (1992) derive the linear dispersion relation

$$D(\omega, k) = -\omega^2(\rho_1+\rho_2)+(\rho_1-\rho_2)gk+\sigma k^3 -\mu_2 \frac{(\mu-1)^2 H^2 k^2}{4\pi\mu(\mu+1)} = 0. \quad (4.4.13)$$

The quantities ω, k are the frequency and wavenumber of the mode creating the disturbance and $\mu = \mu_1/\mu_2$ (at the interface). The critical field strength H_c is found from (4.4.13) as

$$H_c^2 = \frac{8\pi\mu(\mu+1)}{\mu_2(\mu-1)^2} \sqrt{(\rho_1-\rho_2)g\sigma}, \quad (4.4.14)$$

with the critical wavenumber given by

$$k_c^2 = \frac{(\rho_1-\rho_2)g}{\sigma}. \quad (4.4.15)$$

A surface instability (in the form of periodic peaks) sets in for $H \geq H_c$. To study instability along this post bifurcation curve Malik & Singh (1992) develop an analysis with two harmonic modes and set

$$\omega_2 = 2\omega_1 + O(\epsilon), \qquad k_2 = 2k_1 + O(\epsilon).$$

The following expansions are adopted,

$$\eta_2 = \sum_{n=1}^{2} A_n \exp\left(i\theta_n\right) + \text{c.c.},$$

$$\phi_2^{(k)} = (-1)^k \sum_{n=1}^{2} i\left[\frac{\omega_n}{k_n} A_n \exp\left(i\theta_n + (-1)^{k-1}k_n z\right)\right] + \text{c.c.}, \qquad k = 1, 2,$$

$$\psi_2^{(1)} = \frac{H(1-\mu)}{\mu(1+\mu)} \sum_{n=1}^{2} A_n \exp\left(i\theta_n + k_n z\right) + \text{c.c.},$$

$$\psi_2^{(2)} = -\frac{H(1-\mu)}{\mu(1+\mu)} \sum_{n=1}^{2} A_n \exp\left(i\theta_n - k_n z\right) + \text{c.c.},$$

where the notation c.c. stands for complex conjugate, and where the θ_n are given by

$$\theta_n = k_n x_0 - \omega_n t_0, \qquad n = 1, 2.$$

In this manner Malik & Singh (1992) show there is a secondary bifurcation magnetic field strength, H_m. They then use orthogonality of the $O(\epsilon^2)$

solution and $O(\epsilon^3)$ solution to derive a solvability condition and solve the third order problem. Malik & Singh (1992) show that

$$\frac{\partial A_n}{\partial x_1} = 0, \qquad n = 1, 2.$$

The expansion at $O(\epsilon^4)$ is then carried out and by requiring suitable orthogonality with the lower order solutions Malik & Singh (1992) arrive at the following equations for A_1, A_2,

$$\frac{\partial^2 A_1}{\partial t_1^2} + K_1 \frac{\partial A_1}{\partial x_2} = k_m^2 r \left(\frac{H^2}{H_m} - 1 \right) A_1 + Jr A_2 \bar{A}_1 \exp(i\Gamma), \quad (4.4.16)$$

$$\frac{\partial^2 A_2}{\partial t_1^2} + iK_2 \frac{\partial A_2}{\partial x_2} = 4k_m^2 r \left(\frac{H^2}{H_m} - 1 \right) A_2 + Jr A_1^2 \exp(-i\Gamma). \quad (4.4.17)$$

In equations (4.4.16), (4.4.17), K_1, K_2 are coefficients depending on the material parameters at the interface and the associated frequencies and wavenumbers, while the coefficients r, J, and Γ are given by

$$r = \frac{\mu_2 (1 - \mu)^2}{4\pi\mu(1 + \mu)(\rho_1 + \rho_2)},$$

$$\Gamma = \theta_2 - 2\theta_1,$$

$$J = \frac{2k_m^3 (\mu - 1) H^2}{(1 + \mu)}.$$

At this point Malik & Singh (1992) restrict attention to perfect resonance, for which $\Gamma = 0$, and they neglect the variation with x_2. They perform a numerical integration of the resulting ordinary differential equations (4.4.16) and (4.4.17). Their numerical computations are performed for parameter values consistent with those used in experimental situations. The behaviour of $|A_1|$ and $|A_2|$ against time t found by Malik & Singh (1992) strongly suggests $|A_1|$ and $|A_2|$ blow-up in finite time. They finish by writing that such an explosive instability will in practice be inhibited due to dissipation from other effects such as fluid viscosity.

4.5 Temperature Blow-Up in an Ice Sheet

Yuen *et al.* (1986) have developed a model for the temperature at the base of an ice sheet and have analysed the possible consequences of finite time blow-up in the base temperature field, in relation to climatic environmental concerns. Their premise is that heating due to shear stresses could be the cause of an ice flow instability. It is suggested in Yuen *et al.* (1986) and in Schubert & Yuen (1982) that the East Antarctic ice sheet could, in fact, be susceptible to such a shear instability.

A convenient derivation of the equations governing the thermo-mechanical behaviour of an ice sheet or the cold ice section of a glacier may be found in Hutter (1983), section 3.3; pages 145–166 are especially relevant. In that section Hutter (1983) also includes an appraisal of the work of Yuen & Schubert (1979) which is based upon a linearised stability technique. Hutter (1983) deals with equations for two types of ice: temperate ice, which is a mixture of ice and water at $0°C$ (or at the melting temperature T_m $°C$ which depends on pressure), and cold ice, which is ice whose temperature is below the melting temperature.

Hutter (1983) describes cold ice as a non-Newtonian, viscous, heat conducting, incompressible fluid which satisfies the momentum, continuity, and energy balance equations

$$\rho\left(\frac{\partial v_i}{\partial t} + v_j\frac{\partial v_i}{\partial x_j}\right) = -\frac{\partial p}{\partial x_i} + \frac{\partial t'_{ji}}{\partial x_j} + \rho g_i,$$

$$\frac{\partial v_i}{\partial x_i} = 0, \tag{4.5.1}$$

$$\rho\left(\frac{\partial \epsilon}{\partial t} + v_i\frac{\partial \epsilon}{\partial x_i}\right) = t'_{ij}D_{ji} - \frac{\partial q_i}{\partial x_i},$$

where $\rho, v_i, p, D_{ji}, \epsilon, q_i, g_i$ are the density, velocity, pressure, symmetric part of the velocity gradient, internal energy, heat flux, and external force vector, respectively. The tensor t'_{ji} is the deviatoric stress related to the total stress t_{ji} by

$$t_{ij} = t'_{ij} - p\delta_{ij}.$$

The constitutive equations of Hutter (1983) are

$$\rho\left(\frac{\partial \epsilon}{\partial t} + v_i\frac{\partial \epsilon}{\partial x_i}\right) = \rho c_p\left(\frac{\partial T}{\partial t} + v_i\frac{\partial T}{\partial x_i}\right),$$

$$q_i = -\kappa\frac{\partial T}{\partial x_i}, \tag{4.5.2}$$

$$D_{ij} = \mathcal{A}(T)f\left(t'_{II}\right)t'_{ij},$$

where T is the temperature field, c_p is the specific heat at constant pressure, κ is the heat conductivity of ice, $\mathcal{A}(T)$ is a rate factor which typically satisfies an Arrhenius relation of form

$$\mathcal{A}(T) = A\exp\left(\frac{-Q}{kT}\right).$$

Here Q is the activation energy and k is Boltzmann's constant. The quantity

$$t'_{II} = \frac{1}{2}\operatorname{tr}\left(\mathbf{t}'^2\right)$$

is the second stress deviator invariant. The non-Newtonian aspect of the model is evident through the non-linear relationship of D_{ij} depending on t'_{ij}.

These expressions lead to the reduced form of the balance of energy equation which governs the evolutionary behaviour of the temperature field T, namely

$$\rho c_p \left(\frac{\partial T}{\partial t} + v_i \frac{\partial T}{\partial x_i} \right) = \kappa \Delta T + 2 \mathcal{A}(T) f\left(t'_{II}\right) t'_{II}, \qquad (4.5.3)$$

cf. Hutter (1983), equation (3.5). The equations for a parallel sided ice sheet at an angle of γ to the horizontal are then derived by Hutter (1983), equations (3.54). Employment of (4.5.1)–(4.5.3) requires a functional relationship for $f\left(t'_{II}\right)$. This is typically taken to have the form of Glen's flow law, discussed in e.g. Hutter (1983), pp. 83,84. This law has form

$$\dot{\epsilon} = \mathcal{A}(T) \, \text{sgn}(\sigma) |\sigma|^n,$$

where $\dot{\epsilon}$ is the strain rate, σ is the normal stress, and the exponent n is typically $n = 3$, although other values of n may be required, see the discussion on pp. 83,84 of Hutter (1983).

The ice sheet theory of Yuen et al. (1986) essentially employs the equations given above in a one-dimensional model which neglects x-variations. In Yuen et al. (1986) the ice sheet is assumed to be infinite in horizontal extent with an inclination α to the horizontal, with α typically 0.1°. The thickness of the ice sheet is h measured in a direction orthogonal to the sheet, so that the base is at $y = 0$ and the surface at $y = h$. By taking the vertical velocity to be linear in y the mathematical model for the temperature field is then

$$\frac{\partial T}{\partial t} + v_0 \frac{y}{h} \frac{\partial T}{\partial y} = \kappa \frac{\partial^2 T}{\partial y^2} + \frac{2A}{k} \left[\rho g(h - y) \sin \alpha \right]^4 \exp\left(\frac{-E^*}{RT} \right). \qquad (4.5.4)$$

The boundary conditions of Yuen et al. (1986) are that the temperature is known at the free surface of the ice sheet whereas at the base the heat flux is known, so

$$T = T_s \quad \text{at} \quad y = h, \qquad -k \frac{\partial T}{\partial y} = q_b \quad \text{at} \quad y = 0. \qquad (4.5.5)$$

Here q_b ($\approx 41.8 \, \text{mW m}^{-2}$) is the geothermal heat flux at the base of the ice sheet.

Since the normal velocity v is prescribed in (4.5.4), equations (4.5.4) and (4.5.5) are a self-contained system for the temperature field $T(y, t)$ and solution of these will yield $T_b(t)$, the temperature at the base $y = 0$.

The velocity in the direction down the ice sheet, $u(y, t)$, is then determined from the relation

$$\frac{\partial u}{\partial y} = 2A \left[\rho g(h - y) \sin \alpha \right]^3 \exp\left(\frac{-E^*}{RT} \right). \qquad (4.5.6)$$

Equation (4.5.6) is solved subject to the boundary condition

$$u = 0 \qquad \text{at} \quad y = 0. \tag{4.5.7}$$

Yuen *et al.* (1986) observe that the steady solutions to (4.5.4) - (4.5.7) exhibit the possibility of multiple steady states. That is, for $h < h_c$, where h_c may typically be of order 4.7 km, there are two possible values of T_b and surface velocity u_s. These two values depend on what value h has. For $h > h_c$ there are no steady state solutions.

Yuen *et al.* (1986) solve (4.5.4), (4.5.5) by a numerical technique which approximates the spatial derivatives by Hermite cubic splines and integrates in time using the method of lines. Because Yuen *et al.* (1986) are interested in the effect a (relatively) sudden increase in ice sheet thickness will have on the instability, they study the following situation. At time $t = 0$ the ice sheet has thickness h_0. At $t > 0$ the thickness is instantly increased to $h_0 + \Delta h$. The "initial" temperature profile used in the computations is then one which is constant at $T = T_s$ for $h_0 < h < h_0 + \Delta h$, whereas for $h < h_0$ the initial function T is that for the steady distribution with thickness h_0. The accumulation rate is such that at $t = 0$, $v = v_0 y / h_0$, whereas for $t > 0$, $v = v_0 y / (h_0 + \Delta h)$; (in their paper there is a minus sign in front of v_0, but we assume it should be as given here).

The results of Yuen *et al.* (1986) are certainly very interesting. They exhibit computations for various values of the parameters, activation energy E^*, initial height h_0, and incremental height Δh. The critical height h_c depends on E^*. When $h_0 + \Delta h$ is above h_c the computations of Yuen *et al.* (1986) exhibit blow-up in finite time of the temperature, T_b, at the base of the ice sheet. They find the instability to be very sensitive to the value of E^*, and this quantity itself is not known with great accuracy. For example, Yuen *et al.* (1986) note that for temperatures below $-10°C$ E^* could vary between 42 and 84 kJ mol^{-1}. In addition, the increase in ice thickness can dramatically shorten the time to blow-up. For example, for $E^* = 50$ kJ mol^{-1}, with $h_0 = 2$ km and $\Delta h = 8$ km, the time to T_b exploding can be as short as 30 years, whereas with the same activation energy a figure of $\Delta h = 3$ km produces blow-up in T_b on a timescale of around 600 years.

Clearly, finite time blow-up of the base temperature of an ice sheet is very important and can lead to dramatic shear instabilities. The paper of Yuen *et al.* (1986) considers carefully the implications of this blow-up and its relation to possible climate changes and consequential effects on the environment, cf. also Yuen & Schubert (1979), Schubert & Yuen (1982).

The mathematical model employed by Yuen *et al.* (1986) is a very interesting one. As far as I know the only non-linear analysis of this model is by computational means. It would be rewarding to see if the blow-up predictions could be verified by mathematical analysis, and the model as such is well worth further investigation.

5. Blow-Up in Volterra Equations

5.1 Blow-Up for a Solution to a Volterra Equation

There has recently been a systematic investigation of the properties of existence, blow-up, and asymptotic behaviour of a solution to a Volterra equation, or of a solution to a system of such equations. Roberts *et al.* (1993) investigated a single Volterra equation with the underlying spatial domain being the whole of the real line, and the study of the asymptotic behaviour of their solution was extended by Roberts & Olmstead (1996). Olmstead & Roberts (1994) investigate similar problems for a solution to a Volterra equation but in the case where the spatial domain is a bounded region of $\mathbf{R}^1$. The papers of Olmstead and Roberts (1996) and Olmstead *et al.* (1995) consider existence, blow-up and asymptotic behaviour questions for coupled systems of two Volterra equations.

In this chapter we include an exposition of the above work, but first we see why such studies are of interest in mechanics. To do this we briefly describe work of Kapila (1981) on the problem of ignition of a combustible material which can experience a chemical reaction. Further studies of ignition and thermal explosions in such combustible materials are contained in the papers of Olmstead (1983), and Glenn Lasseigne & Olmstead (1983, 1991). An introductory account of combustion phenomena with an Arrhenius law for the chemical reaction is given by Logan (1994), pp. 238–240. The effect of coupled fluid motion in a porous medium together with a chemical reaction of Arrhenius type is described by Nield & Bejan (1992), see pp. 34, 35 and 286. Also, the effect of chemical reactions and convective motion in a viscous fluid is reported in Chap. 13 of the book by Straughan (1992).

Kapila (1981) reports theoretical work involving a rigid solid which is combustible, confined in a suitable container, and which can undergo a chemical reaction of the Arrhenius type. It is reported there that such a material can burst into flame by two mechanisms. One is by thermal explosion incurred by self-induced internal heating. The other is due to an external ignition (e.g. heat flux) supply. Kapila's (1981) paper concentrates on the latter although references to the former are also given.

Kapila (1981) assumes the material occupies the half space $x > 0$ and restricts attention to disturbances in one space dimension. The starting position for the physical problem is one for which the material has fixed temperature and fixed fuel mass fraction. The non-dimensionless form of the partial differential equations for the behaviour of the temperature, $\theta(x,t)$, and the fuel mass fraction, $Y(x,t)$, with an Arrhenius chemical reaction are given by Kapila (1981) as

$$
\begin{aligned}
\frac{\partial \theta}{\partial t} &= \frac{\partial^2 \theta}{\partial x^2} + AY \exp\left(-\frac{E}{\theta}\right), \\
\frac{\partial Y}{\partial t} &= \frac{1}{L}\frac{\partial^2 Y}{\partial x^2} - \frac{A}{B}Y \exp\left(-\frac{E}{\theta}\right),
\end{aligned}
\tag{5.1.1}
$$

where A, B, L and E are non-dimensional constants, A, B and E being related to the chemical reaction. The boundary conditions considered by Kapila (1981) are

$$
\begin{aligned}
\frac{\partial \theta}{\partial x}(0,t) &= -1, & \theta(\infty,t) &= 1, \\
\frac{\partial Y}{\partial x}(0,t) &= 0, & Y(\infty,t) &= 1,
\end{aligned}
\tag{5.1.2}
$$

while the initial conditions are

$$
\theta(x,0) = 1, \qquad Y(x,0) = 1. \tag{5.1.3}
$$

By using dimensional arguments involving the respective sizes of the coefficients A, B, L and E, Kapila (1981) shows that the variable Y may be removed from consideration in the above boundary initial value problem. Then in terms of the modified temperature ϕ given by

$$
\phi = \theta - \theta_I,
$$

where θ_I is the temperature in an inert stage, Kapila (1981) shows that the relevant boundary initial value problem to be solved is

$$
\begin{aligned}
\frac{\partial \phi}{\partial t} &= \frac{\partial^2 \phi}{\partial x^2} + \frac{\bar{A}}{B}(B - \phi)\exp\left\{\frac{\theta_c^2}{\epsilon}\left[\frac{1}{\theta_c} - \frac{1}{(\phi + \theta_I)}\right]\right\}, \\
\frac{\partial \phi}{\partial x}(0,t) &= 0, \qquad \phi(\infty,t) = 0, \\
\phi(x,0) &= 0.
\end{aligned}
\tag{5.1.4}
$$

The coefficients $\bar{A}, \theta_c$ and ϵ are constants.

Kapila (1981) shows that the thermal ignition process due to the heat flux begins with an inert stage, and this is followed by a transistion stage. By using an asymptotic analysis, Kapila (1981) scales space and time and in scaled coordinates X, τ, he then expands the variable $\phi(X,\tau)$ as

$$
\phi = \epsilon \psi_0(X,\tau) + \epsilon^{3/2}\psi_1(X,\tau) + \dots, \tag{5.1.5}
$$

for an expansion parameter ϵ. The boundary initial value problems for the functions ψ_0 and ψ_1 are found to be

$$\frac{\partial \psi_0}{\partial \tau} = \frac{\partial^2 \psi_0}{\partial X^2}, \qquad \tau > -\infty, \ X > 0,$$

$$\psi_0(0,\tau) = f_0(\tau), \qquad \frac{\partial \psi_0}{\partial X}(0,\tau) = -A_0 P(\tau), \qquad \psi_0(\infty,\tau) = 0, \qquad (5.1.6)$$

$$\psi_0(X,-\infty) = 0,$$

and

$$\frac{\partial \psi_1}{\partial \tau} = \frac{\partial^2 \psi_1}{\partial X^2}, \qquad \tau > -\infty, \ X > 0,$$

$$\psi_1(0,\tau) = f_1(\tau),$$

$$\frac{\partial \psi_1}{\partial X}(0,\tau) = \frac{3}{2} A_0^2 [P(\tau)]^2 - A_0 P(\tau) f_1(\tau), \qquad (5.1.7)$$

$$\psi_1(\infty,\tau) = 0,$$

$$\psi_1(X,-\infty) = 0.$$

In these equations A_0 is a constant, and the function P is defined by

$$P(\tau) = \exp\left[f_0(\tau) + \frac{1}{\sqrt{\pi t_c}} \tau \right],$$

with f_0, f_1 functions to be satisfied subject to the conditions

$$f_0(-\infty) = 0, \qquad f_1(-\infty) = 0.$$

The Laplace transform is taken of the equation for ψ_0 and from the equation which results by this procedure, Kapila (1981) shows that $\bar{f}_0(\bar{\tau})$, given by

$$\bar{f}_0(\bar{\tau}) = f_0(\tau),$$

with $\bar{\tau}$ defined by

$$\bar{\tau} = \log\left[(\pi t_c)^{1/4} A_0 \right] + \frac{\tau}{\sqrt{\pi t_c}},$$

satisfies the Volterra equation

$$\bar{f}_0(\bar{\tau}) = \frac{1}{\sqrt{\pi}} \int_{-\infty}^{\bar{\tau}} \frac{1}{\sqrt{\bar{\tau} - s}} \exp\left[\bar{f}_0(s) + s \right] ds. \qquad (5.1.8)$$

The paper of Kapila (1981) continues the solution of the thermal ignition problem by showing how to solve (5.1.8) asymptotically for $f_0(\tau)$ and hence obtain $\psi_0(X,0)$ for small X. A similar procedure is employed to find $f_1(\tau)$. Kapila (1981) shows how this leads to a thermal runaway. In fact, Kapila (1981) develops the analysis much further, analysing the ignition stage of burning in detail, and then shows how this develops into an explosion stage, eventually explaining how a propagating thermal wave develops.

Other cases from combustion theory which lead to non-linear Volterra equations not dissimilar to (5.1.8) are considered by Olmstead (1983) and Glenn Lasseigne & Olmstead (1983, 1991). These are a partial justification for a study of blow-up in a general Volterra equation. Olmstead *et al.* (1994) study the formation of shear bands in steel which has been subjected to a very high rate of strain. They show that this formation is accompanied by a very rapid increase in temperature, and a typical equation for the temperature, $u(t)$, in that case is

$$u(t) = \frac{\gamma}{\sqrt{\pi}} \int_0^t \frac{(1+s)^q [u(s)+1]^p}{\sqrt{t-s}} \, ds. \tag{5.1.9}$$

5.1.1 A General Non-linear Volterra Equation

Motivated by the derivation of equations like (5.1.8) and (5.1.9) in differing areas of mechanics, Roberts *et al.* (1993) begin a careful investigation into the properties of solutions to the Volterra equation

$$u(t) = \int_{t_0}^t k(t-s) G[u(s), s] \, ds, \qquad \text{for } t \geq t_0, \tag{5.1.10}$$

for non-linearities of the form

$$G[u(t), t] = r(t) \, g[u(t) + h(t)]. \tag{5.1.11}$$

The functions r and h are given and based on those found in mechanics. The non-linear function g likewise satisfies conditions typical of those found in real situations. In particular, g, r and h are required to satisfy

$$g(u) > 0, \quad g'(u) > 0, \quad g''(u) > 0, \qquad \text{for } u > 0,$$
$$r(t) > 0, \quad r'(t) \geq 0, \qquad \text{for } t \geq t_0,$$

and

$$0 < h_0 \leq h(t) \leq h_\infty < \infty, \quad h'(t) \geq 0, \qquad \text{for } t \geq t_0.$$

The kernel, $k(t-s)$, satisfies the inequalities

$$k(t-s) \geq 0, \quad k'(t-s) < 0, \qquad \text{for } s \in [t_0, t).$$

Another motivation for studying the general non-linear Volterra equation (5.1.10) is provided by Roberts *et al.* (1993). They show that the solution to the following boundary initial value problem involving a non-linear point source term,

$$\begin{aligned} &\frac{\partial v}{\partial t} = \frac{\partial^2 v}{\partial x^2} + 2\delta(x) g[v(x,t)], \qquad x \in \mathbf{R}, \, t > 0, \\ &v(-\infty, t) = 0, \qquad v(+\infty, t) = 0, \\ &v(x, 0) = v_0(x), \end{aligned} \tag{5.1.12}$$

can be related to a study of the Volterra equation

$$u(t) = \frac{1}{\sqrt{\pi}} \int_0^t \frac{g[u(s) + h(s)]}{\sqrt{t-s}} \, ds \,,$$

in which

$$h(t) = v(0,t) - u(t)$$

$$= \frac{1}{\sqrt{\pi t}} \int_{-\infty}^{\infty} e^{-x^2/4t} v_0(x) \, dx.$$

A lucid introduction to Volterra equations and how one transforms a differential equation into a Volterra equation and vice-versa is given by Logan (1987), pp. 199–223.

The work of Roberts $et\ al.$ (1993) on the general non-linear Volterra equation (5.1.10) uses the function $I(t)$ defined by

$$I(t) = \int_{t_0}^t k(t-s) r(s) \, ds, \qquad t \geq t_0, \tag{5.1.13}$$

and the number κ given by

$$\kappa = \int_{h_0}^{\infty} \frac{d\eta}{g(\eta)} \,. \tag{5.1.14}$$

Roberts $et\ al.$ (1993) first show that any continuous solution to (5.1.10) which is differentiable, must be positive and increasing, for $t > t_0$. They also prove the following theorem which establishes existence of a unique, continuous solution to (5.1.10) with the property that

$$0 \leq u(t) \leq M < \infty, \qquad \text{for } t \in [t_0, t^*), \tag{5.1.15}$$

for a constant M defined to satisfy the theorem below.

Theorem. (Roberts, Glenn Lasseigne & Olmstead (1993)).
There is a unique continuous solution to equation (5.1.10) such that restrictions (5.1.15) hold for any $M < M^*$. The number M^* is defined as the smallest solution to the equation

$$\frac{M^*}{g(M^* + h_\infty)} = \frac{1}{g'(M^* + h_\infty)} \,. \tag{5.1.16}$$

Roberts $et\ al.$ (1993) then use an elegant contradiction argument to deduce that if

$$I(t^{**}) = \kappa = \int_{h_0}^{\infty} \frac{d\eta}{g(\eta)} < \infty,$$

equation (5.1.10) cannot have a continuous solution for $t \geq t^{**}$. By combining this result with their previous theorem, Roberts $et\ al.$ (1993) are able to produce the following blow-up result.

Theorem. (Roberts, Glenn Lasseigne & Olmstead (1993)).
If the number M^* defined in (5.1.16) and the number κ defined in (5.1.14) are such that the values of $M^*/g(M^* + h_\infty)$ and $\kappa(< \infty)$ are both in the range of the function $I(t)$, then the unique positive, continuous and increasing solution to equation (5.1.10) must cease to exist at a time $t = \hat{t}$. The number $\hat{t}$ satisfies the estimates

$$t_0 < t^* \leq \hat{t} \leq t^{**} < \infty.$$

The lower and upper bounds t^* and t^{**} are given by

$$t^* = \frac{1}{I\big(M^*/g(M^* + h_\infty)\big)},$$

and

$$t^{**} = \frac{1}{I(\kappa)} \, .$$

Roberts et (1993) apply their theorem given above to produce precise conditions for blow-up for a solution to the Volterra equations

$$u(t) = \frac{\gamma}{\sqrt{\pi}} \int_{-\infty}^{t} \frac{e^{\alpha s - c}}{\sqrt{t - s}} \left[e^{u(s) + c} \right] ds, \tag{5.1.17}$$

and

$$u(t) = \int_{-\infty}^{t} k(t - s) e^{s - c} \left[e^{u(s) + c} \right] ds, \tag{5.1.18}$$

where the kernel in (5.1.18) is given by

$$k(t) = \gamma \left(\frac{1}{\sqrt{\pi t}} - e^t \operatorname{erfc} \sqrt{t} \right),$$

and also for the Volterra equation

$$u(t) = \frac{\gamma}{\sqrt{\pi}} \int_0^t \frac{[u(s) + 1]^p}{\sqrt{t - s}} \, ds. \tag{5.1.19}$$

The first two examples, (5.1.17) and (5.1.18), are motivated by the thermal combustion explosion problems, while the last, equation (5.1.19), has its motivation in the analysis of the formation of shear bands in steel, as studied by Olmstead *et al.* (1994).

Roberts *et al.* (1993) also give examples where blow-up does *not* occur. They explore the connection with a parabolic equation with a point source non-linearity, and also investigate the asymptotic behaviour of the solution in the vicinity of the blow-up time in some detail.

The work of Roberts & Olmstead (1996) continues the asymptotic development part of the above work in greater detail and also extends the class of kernels $k(t - s)$ and non-linearities $g(u)$. The mathematical analysis employs complex function theory, with some elegant asymptotic analysis involving Mellin transforms.

5.1.2 Volterra Equations Motivated by Partial Differential Equations on a Bounded Spatial Domain

The papers of Olmstead & Roberts (1994, 1996) investigate similar properties of solutions to Volterra equations to those just outlined. However, the partial differential equations out of which the Volterra equations arise are defined for a bounded spatial domain and hence the mathematical analysis is necessarily different.

Olmstead & Roberts (1994) investigate Volterra equations arising from the non-linear partial differential equation

$$\frac{\partial v}{\partial t} = \frac{\partial^2 v}{\partial x^2} + F\big(v(x,t),x\big), \qquad x \in (0,\ell), \, t > 0, \tag{5.1.20}$$

in which the non-linear function F has form

$$F\big(v(x,t),x\big) = \delta(x-a)g\big(v(x,t)\big), \qquad \text{some } a \in (0,\ell), \tag{5.1.21}$$

$\delta(x)$ being the Dirac delta function. The physical application they attribute this partial differential equation to is where a combustible material is ignited by a heated fine wire, or by small electrodes. In either case the ignition source is created by the supply of a relatively large quantity of energy delivered to a small area. The representation of this by a delta function is deemed appropriate. Other studies of equations (5.1.20), (5.1.21) are by Chadam *et al.* (1992), and by Chadam & Yin (1993). The latter have $g\big(v(a,t)\big)$ in (5.1.21).

Olmstead & Roberts (1994) examine equations (5.1.20) and (5.1.21) subject to the solution satisfying the initial condition

$$v(x,0) = v_0(x), \qquad x \in [0,\ell], \tag{5.1.22}$$

and either boundary conditions of Neumann type,

$$\frac{\partial v}{\partial x}(0,t) = 0, \qquad \frac{\partial v}{\partial x}(\ell,t) = 0, \tag{5.1.23}$$

or boundary conditions of Dirichlet type

$$v(0,t) = 0, \qquad v(\ell,t) = 0. \tag{5.1.24}$$

Olmstead & Roberts (1994) transform (5.1.20) to a Volterra equation of form

$$v(x,t) = \int_0^t \int_0^\ell G(x,t|\xi,s) \, F\big(v(\xi,s),\xi\big) \, d\xi \, ds$$
$$+ \int_0^\ell G(x,t|\xi,0) \, v_0(\xi) \, d\xi. \tag{5.1.25}$$

The Green's functions G_N and G_D for the Neumann and Dirichlet problems are given explicitly in Olmstead & Roberts (1994) in terms of a spectral representation and in terms of a series involving exponentials as this is essential to their analysis of the Volterra equation (5.1.25). By proceeding in this manner Olmstead & Roberts (1994) are able to write both the Neumann and Dirichlet problems as Volterra equations of form

$$u(t) = \int_0^t k(t - s)g\big(u(s) + h(s)\big) \, ds, \qquad t \geq 0, \tag{5.1.26}$$

where the explicit representations for the kernels, k, and functions, h, are derived in Olmstead & Roberts (1994). By employing a contraction mapping argument, they show that a continuous and unique solution to (5.1.26) exists for $t \in [0, t^*)$. They also produce a method for estimating the time t^*.

Conditions for finite time blow-up of a solution to (5.1.26) are also derived by Olmstead & Roberts (1994). The asymptotic behaviour of the growth rate in the neighbourhood of the blow-up time is studied in detail using complex function arguments and the Mellin transform. The analysis is quite elegant and the findings are very interesting. For the Neumann problem it is found that blow-up always occurs. However, blow-up in the Dirichlet problem does *not* occur if $a(\ell - a)/\ell$ is sufficiently small. This means that blow-up will not occur if the source term is located sufficiently close to either edge of the strip. Bounds for the blow-up time are derived and it is shown that the asymptotic growth rate of the solution in the blow-up case is the same for both the Dirichlet and Neumann problems.

Olmstead and his co-workers have produced an interesting collection of results regarding existence, non-existence by blow-up in finite time, investigating in detail the behaviour of the solution near the blow-up time, for a variety of Volterra equations arising in mechanics. In very recent work they extend this to the situation where the mechanics problems require the solution to a *system* of Volterra equations rather than a single equation.

5.2 Blow-Up for a Solution to a System of Volterra Equations

Olmstead & Roberts (1996) extend their earlier analysis, Olmstead & Roberts (1994), to the non-linear partial differential equation

$$\frac{\partial v}{\partial t} = \frac{\partial^2 v}{\partial x^2} + \delta(x - a)g\big(v(x, t)\big) \|v(t)\|_{L^1(0,\ell)}^r, \tag{5.2.1}$$

where $x \in (0, \ell), t > 0$, and $a \in (0, \ell)$. The non-linearity is non-local due to the presence of the $L^1(0, \ell)$ term. This is an interesting non-linearity which involves a non-local term concentrated at the spatial position $x = a$. Again

Olmstead & Roberts (1996) study both Neumann and Dirichlet boundary conditions. They transform the boundary initial value problem for the partial differential equation (5.2.1) by converting to a Volterra equation. However, in this situation, a single Volterra equation is not sufficient. If the notation $V(t)$ is used to denote the $L^1(0, \ell)$ norm of v, i.e.

$$V(t) = \|v(t)\|_{L^1(0,\ell)},$$

then they show the solution to the partial differential equation (5.2.1) must satisfy the pair of Volterra equations

$$
\begin{aligned}
u(t) &= \int_0^t k(t-s)g\big(u(s) + h(s)\big)\big[\tilde{u}(s) + \tilde{h}(s)\big]^r \, ds, & t &\geq 0, \\
\tilde{u}(t) &= \int_0^t \tilde{k}(t-s)g\big(u(s) + h(s)\big)\big[\tilde{u}(s) + \tilde{h}(s)\big]^r \, ds, & t &\geq 0,
\end{aligned}
\tag{5.2.2}
$$

where $\tilde{u}(t) = V(t) - \|h(t)\|_{L^1(0,\ell)}$. The transformation of the partial differential equation thus yields an interesting system of Volterra equations with different kernels k and $\tilde{k}$. The paper of Olmstead & Roberts (1996) establishes existence of a solution, investigates conditions for finite time blow-up of a solution, and discovers that blow-up is always evident in the Neumann problem whereas it may or may not occur for the Dirichlet problem. A specific example involving a power law non-linearity is studied in detail and estimates for the blow-up time and asymptotic behaviour near blow-up are calculated precisely.

5.2.1 Coupled Non-linear Volterra Equations Which May Arise from Non-linear Parabolic Systems

Olmstead *et al.* (1995) consider a coupled system of Volterra equations. They introduce their investigation by looking at coupled parabolic equations which arise in mechanics. In particular, they study the solution to the following two boundary initial value problems for coupled non-linear systems of parabolic partial differential equations.

The first boundary initial value problem is

$$
\begin{aligned}
\frac{\partial w_1}{\partial t} &= \frac{\partial^2 w_1}{\partial x^2} + \delta(x-\ell)F_1\big(w_2(x,t)\big), & x &\in (0,L), \, t > 0, \\
\frac{\partial w_2}{\partial t} &= \frac{\partial^2 w_2}{\partial x^2} + \delta(x-\ell)F_2\big(w_1(x,t)\big), & x &\in (0,L), \, t > 0, \\
\frac{\partial w_\alpha}{\partial x}(0,t) &= 0, \qquad \frac{\partial w_\alpha}{\partial x}(L,t) = 0, & \alpha &= 1,2, \, t \geq 0, \\
w_1(x,0) &= w_1^0(x), \qquad w_2(x,0) = w_2^0(x), & x &\in [0,L],
\end{aligned}
\tag{5.2.3}
$$

while the second has the non-linearity in the boundary condition, and is

$$\frac{\partial w_1}{\partial t} = \frac{\partial^2 w_1}{\partial x^2}, \qquad x \in (0, L), \ t > 0,$$

$$\frac{\partial w_2}{\partial t} = \frac{\partial^2 w_2}{\partial x^2}, \qquad x \in (0, L), \ t > 0,$$

$$\frac{\partial w_\alpha}{\partial x}(0, t) = 0, \qquad \alpha = 1, 2, \ t \geq 0,$$

$$\frac{\partial w_1}{\partial x}(L, t) = F_1\big(w_2(L, t)\big), \qquad t \geq 0, \qquad\qquad (5.2.4)$$

$$\frac{\partial w_2}{\partial x}(L, t) = F_2\big(w_1(L, t)\big), \qquad t \geq 0,$$

$$w_1(x, 0) = w_1^0(x), \qquad w_2(x, 0) = w_2^0(x), \qquad x \in [0, L].$$

Both systems (5.2.3) and (5.2.4) are shown by Olmstead et $al.$ (1995) to be convertible into a coupled system of non-linear Volterra equations of form

$$u_1(t) = \int_0^t k(t - s) F_1\big(u_2(s) + h_2(s)\big)\, ds, \qquad t \geq 0,$$
$$\qquad\qquad (5.2.5)$$
$$u_2(t) = \int_0^t k(t - s) F_2\big(u_1(s) + h_1(s)\big)\, ds, \qquad t \geq 0.$$

The explicit Green's function $G(x, t | \xi, s)$ involved in recasting (5.2.3) or (5.2.4) in a form like (5.2.5) is given in Olmstead et $al.$ (1995).

The paper of Olmstead et $al.$ (1995) first establishes an existence theorem for a solution to (5.2.5) and this is useful in the continuation arguments necessary in the blow-up results which subsequently are derived by the same writers.

Olmstead et $al.$ (1995) use a comparison argument to demonstrate that blow-up in finite time of a solution to (5.2.5) is to be expected for a class of non-linearities and functions h_1, h_2. They show that if either u_1 or u_2 ceases to exist at some time t_c, then the other component does likewise.

In Olmstead et $al.$ (1995) specific applications of the general results are detailed for the system of Volterra equations

$$u_1(t) = \int_0^t k(t - s)\big(u_2(s) + h_2(s)\big)^{p_1}\, ds,$$
$$\qquad\qquad (5.2.6)$$
$$u_2(t) = \int_0^t k(t - s)\big(u_1(s) + h_1(s)\big)^{p_2}\, ds,$$

for $p_1, p_2 > 0$, and for the system

$$u_1(t) = \int_0^t k(t - s) e^{\gamma_1 \big(u_2(s) + h_2(s)\big)}\, ds,$$
$$\qquad\qquad (5.2.7)$$
$$u_2(t) = \int_0^t k(t - s) e^{\gamma_2 \big(u_1(s) + h_1(s)\big)}\, ds.$$

Explicit details of blow-up are produced for a solution to each of systems (5.2.6) and (5.2.7). The paper of Olmstead *et al.* (1995) is completed by deriving asymptotic estimates for the solution behaviour in the vicinity of the blow-up time, t_c. In fact, the solution growth for the power-law non-linear system (5.2.6) is of asymptotic form

$$u_1(t) \sim \frac{A_1}{(t_c - t)^{\ell_1}}, \qquad \text{as } t \to t_c,$$

$$u_2(t) \sim \frac{A_2}{(t_c - t)^{\ell_2}}, \qquad \text{as } t \to t_c,$$

where the powers ℓ_1 and ℓ_2 are given by

$$\ell_1 = \frac{p_1 + 1}{2(p_1 p_2 - 1)}, \qquad \ell_2 = \frac{p_2 + 1}{2(p_1 p_2 - 1)},$$

and where the constants A_1 and A_2 are defined by

$$A_1 = \left\{ \frac{2^{p_1+1} \Gamma(p_1 \ell_2)}{\Gamma(p_1 \ell_2 - 1/2)} \left[\frac{\Gamma(p_2 \ell_1)}{\Gamma(p_2 \ell_1 - 1/2)} \right]^{p_1} \right\}^{1/(p_1 p_2 - 1)},$$

$$A_2 = \left\{ \frac{2^{p_2+1} \Gamma(p_2 \ell_1)}{\Gamma(p_2 \ell_1 - 1/2)} \left[\frac{\Gamma(p_1 \ell_2)}{\Gamma(p_1 \ell_2 - 1/2)} \right]^{p_2} \right\}^{1/(p_1 p_2 - 1)},$$

with $\Gamma(\cdot)$ being the gamma function. The behaviour of a solution to the system (5.2.7) with the exponential non-linearities is shown in Olmstead *et al.* (1995) to be of form

$$u_1(t) \sim \frac{1}{2\gamma_2} \log\left(\frac{1}{t_c - t} \right), \qquad \text{as } t \to t_c,$$

$$u_2(t) \sim \frac{1}{2\gamma_1} \log\left(\frac{1}{t_c - t} \right), \qquad \text{as } t \to t_c.$$

In the next chapter we move away from mechanics to another very interesting example of blow-up in finite time. This is seen to be in the area of mathematical biology, although as we explain, there are many mathematical similarities in the derivation of equations in the mathematical biology model and those found in mechanics. Before commencing this, however, we add that the work of Olmstead and his co-workers outlined here is very interesting and should find application in many other practical areas of applied mathematics.

6. Chemotaxis

6.1 Mathematical Theories of Chemotaxis

Chemotaxis is a term which occurs frequently in biology. It refers to a phenomenon of chemically directed movement. It is instructive to consider a population of a biological species. In a diffusion process, if the species is densely crowded together then diffusion acts to spread the population outward in space. Chemotaxis is a completely different effect whereby the species is attracted towards a high chemical concentration. For example, as Murray (1990), p. 241, points out, female members of certain species exude a chemical to attract male members of the same species.

A much studied chemotaxis process in mathematical biology is that of the formation of amoebae into a slime mold. A mathematical theory for this was developed by Keller & Segel (1970), and an account may be found in Lin & Segel (1974), p. 22, see also Murray (1990), p. 242. Models for chemotaxis were also given by Keller & Segel (1971a) and the same writers studied the effects of travelling waves in their theory in Keller & Segel (1971b).

The basic biological process of the slime mold amoebae is given in Lin & Segel (1974), p. 22. They describe that the amoebae feed on bacteria and when their food supply is abundant they propagate by division into two. Whenever, however, the food supply becomes scarce, an interphase period begins in which the amoebae move in a weak and random manner. In this phase the amoebae are effectively spread evenly over their environment. After some hours the amoebae begin to group together in a striking manner and form a many celled slug which can contain approximately 200 000 cells. This slug sometime later stops moving around and erects a stalk which contains spores. These spores eventually re-emerge as amoebae and the life cycle of the slime mold amoebae begins afresh.

To describe the above process in terms of a mathematical model, Keller & Segel (1970) use a coupled system of reaction–diffusion partial differential equations into which is incorporated the chemotaxis effect. The general procedure behind diffusion models in biology is aptly explained in Murray (1990), Chap. 9. Basically, the idea is that one writes equations for functions c_i, $i = 1, \ldots, N$, say, which represent a species population(s), and possibly chemicals they feed on, absorb, or emit. The general form of a diffusion equation in a domain Ω (in biology the spatial domain Ω is usually a subset of

$\mathbf{R}^2$ or $\mathbf{R}^1$), is

$$\frac{\partial}{\partial t} \int_{\Omega} c_i(\mathbf{x}, t)\, dx = -\int_{\Gamma} \mathbf{J}^{(i)} \cdot \mathbf{n}\, dA + \int_{\Omega} f_i\, dx. \qquad (6.1.1)$$

Here Γ is the boundary of Ω, $\mathbf{n}$ is the unit outward normal to Γ, $\mathbf{J}^{(i)}$ is a flux vector associated with the ith equation, and f_i is a source term. By using the divergence theorem to write (6.1.1) as an equation over Ω and reducing to point form, this equation assumes the form

$$\frac{\partial c_i}{\partial t} + \nabla \cdot \mathbf{J}^{(i)} = f_i, \qquad i = 1, \ldots, N, \qquad (6.1.2)$$

holding typically in $\Omega \times (0, \infty)$. Such a procedure of deriving diffusion equations is familiar in continuum mechanics. In the same spirit as is necessary in continuum mechanics, in mathematical biology one needs to specify the domain Ω, the conditions on Γ, and the fluxes $\mathbf{J}^{(i)}$ and source terms f_i. This is where an intimate knowledge of the biological process is essential. Murray (1990), in his preface writes *"mathematical biology is ... the most exciting modern application of mathematics ... the use of esoteric mathematics arrogantly applied to biological problems by mathematicians who know little about the real biology, together with unsubstantiated claims as to how important such theories are, does little to promote the interdisciplinary involvement which is so essential."*

To progress from (6.1.2) Keller & Segel (1970) derive suitable forms for the fluxes and source terms in the slime mold process, based entirely on what happens biologically. They begin with a model with four constituents c_i which they denote by $a(x, y, t)$, the concentration of amoebae, $\rho(x, y, t)$, the concentration of acrasin, $\eta(x, y, t)$, the concentration of acrasinase, and $c(x, y, t)$, the concentration of a chemical produced by a reaction between the acrasin and acrasinase. The acrasin is a chemical produced by the amoebae which is responsible for chemotaxis and the aggregation effect. The acrasin is degraded by an enzyme known as acrasinase. Based on these facts, Keller & Segel (1970) develop a theory for slime mold amoebae aggregation, based on the partial differential equations (6.1.2) employing the dependent constituents a, ρ, η and c as the functions c_i. The flux terms of Keller & Segel (1970) have form

$$\begin{aligned}
\mathbf{J}^{(a)} &= D_1 \nabla \rho - D_2 \nabla a, \\
\mathbf{J}^{(\rho)} &= -D_\rho \nabla \rho, \\
\mathbf{J}^{(c)} &= -D_c \nabla c, \\
\mathbf{J}^{(\eta)} &= -D_\eta \nabla \eta,
\end{aligned} \qquad (6.1.3)$$

where the D_2, D_ρ, D_c and D_η coefficients represent diffusion of their associated species, while it is the D_1 term which represents chemotaxis. Note that we have followed Keller & Segel's (1970) convention of using $\mathbf{J}^{(a)}$ for the flux associated with the equation for a, rather than $\mathbf{J}^{(1)}$, etc. The source terms are

$$f^{(a)} = 0,$$
$$f^{(\rho)} = -k_1 \rho \eta + k_{-1} c + a f(\rho),$$
$$f^{(c)} = k_1 \rho \eta - (k_{-1} + k_2) c,$$
$$f^{(\eta)} = -k_1 \rho \eta + (k_{-1} + k_2) c + a g(\rho, \eta),$$

$$(6.1.4)$$

where f and g must be further specified. Equations (6.1.4) are postulated by Keller & Segel (1970) for good biological reasons. It is not the purpose of this monograph to go into profound detail of these aspects, although we believe an understanding of the theory behind the partial differential equations which eventually arise is very helpful. A good account of the motivation for the fluxes and source terms is given by Murray (1990), p. 242, although he restricts attention to the simplified Keller–Segel (1970) model outlined below. This simplified model contains two partial differential equations for the quantities a and ρ, with the c and η effects being eliminated.

The final four partial differential equations derived by Keller & Segel (1970) arise from (6.1.2)–(6.1.4). They are

$$\frac{\partial a}{\partial t} = -\nabla(D_1 \nabla \rho) + \nabla(D_2 \nabla a),$$
$$\frac{\partial \rho}{\partial t} = D_\rho \Delta \rho - k_1 \rho \eta + k_{-1} c + a f(\rho),$$
$$\frac{\partial c}{\partial t} = D_c \Delta c + k_1 \rho \eta - (k_{-1} + k_2) c,$$
$$\frac{\partial \eta}{\partial t} = D_\eta \Delta \eta - k_1 \rho \eta + (k_{-1} + k_2) c + a g(\rho, \eta),$$

$$(6.1.5)$$

where the diffusion coefficient D_2 and the chemotaxis coefficient D_1 may depend on the dependent variables.

6.1.1 A Simplified Model

Most research based on the Keller & Segel (1970) theory would appear to have focussed on use of a simplified theory which uses equations only for the concentrations of amoebae and acrasin. To derive this model Keller & Segel (1970) make assumptions regarding the chemical state of the process and so assume relations between the variables and the reaction coefficients k_1, k_{-1} and k_2. In this manner they reduce equations (6.1.5) to the following coupled system of partial differential equations

$$\frac{\partial a}{\partial t} = -\nabla(D_1 \nabla \rho) + \nabla(D_2 \nabla a),$$
$$\frac{\partial \rho}{\partial t} = D_\rho \Delta \rho - k(\rho) \rho + a f(\rho).$$

$$(6.1.6)$$

Keller & Segel (1970) adopt the form

$$k(\rho) = \frac{\eta_0 k_2 K}{1 + K\rho},$$

for η_0, K constants. Murray (1990), p. 243, uses D_2 constant and gives various forms for the chemotaxis coefficient D_1, such as

$$D_1 = \chi_0 a, \qquad D_1 = \frac{\chi_0 a}{\rho}, \qquad \text{or} \qquad D_1 = \frac{\chi_0 K a}{(K + \rho)^2},$$

where χ_0 and K are positive constants.

Although the slime mold amoebae process is a beautiful one, our interest in this monograph centres on blow-up results for solutions to partial differential equations and we henceforth turn to two recent studies which have found such behaviour in ramifications of the simplified Keller & Segel (1970) theory, as described by the partial differential equations (6.1.6).

6.2 Blow-Up in Chemotaxis When There Are Two Diffusion Terms

Jäger & Luckhaus (1992) treat $D_2, k(\rho)$ and $f(\rho)$ as constants and write system (6.1.6) in the form

$$
\begin{aligned}
\frac{\partial a}{\partial t} &= \Delta a - \chi \nabla(a \nabla \rho), \\
\frac{\partial \rho}{\partial t} &= \gamma \Delta \rho - \mu \rho + \beta a,
\end{aligned}
\tag{6.2.1}
$$

where χ, γ, μ and β are positive constants. They assume Ω is a bounded region in $\mathbf{R}^2$, and the functions a, ρ are subject to the boundary conditions

$$\frac{\partial a}{\partial \nu} = 0, \qquad \frac{\partial \rho}{\partial \nu} = 0, \qquad \text{on } \Gamma, \tag{6.2.2}$$

where $\partial/\partial \nu$ denotes the outward normal derivative. The solutions must satisfy the initial data

$$a(\mathbf{x}, 0) = a_0(\mathbf{x}), \qquad \rho(\mathbf{x}, 0) = \rho_0(\mathbf{x}), \tag{6.2.3}$$

where the prescribed data functions a_0 and ρ_0 are such that $a_0, \rho_0 \geq 0$ in Ω.

Jäger & Luckhaus (1992) deal with an asymptotic form of the system of partial differential equations (6.2.1) derived by assuming the diffusion coefficient γ is large, the coefficient β has form $\beta = \alpha \gamma$, with α of order one, while the coefficient μ is likewise of order one. By integrating each of the partial differential equations in (6.2.1) over Ω they show that

$$\bar{a}(t) = \bar{a}_0,$$

$$\frac{1}{\gamma}\left(\frac{\partial}{\partial t} + \mu\right)\bar{\rho} = \alpha\bar{a} = \alpha\bar{a}_0, \tag{6.2.4}$$

where $\bar{f}$ denotes an average of f over Ω, in the sense that

$$\bar{f} = \frac{1}{M(\Omega)}\int_\Omega f\,dx.$$

Equations (6.2.4) follow from (6.2.1) due to the Neumann boundary conditions (6.2.2). Since $\int_\Omega \Delta\rho\,dx = 0$, one may subtract the averaged form of $(6.2.1)_2$ from $(6.2.1)_2$ itself to find,

$$\frac{1}{\gamma}\left[\frac{\partial}{\partial t}(\rho - \bar{\rho}) + \mu(\rho - \bar{\rho})\right] = \Delta\rho + \alpha a - \alpha\bar{a}_0\,.$$

Jäger & Luckhaus (1992) use this equation in the limit $\gamma \to \infty$, and then they rescale the resulting system in terms of $\bar{a}_0$ to derive a modified system of partial differential equations for the variables a and ρ. The new system is

$$\frac{\partial a}{\partial t} = \Delta a - \chi\nabla(a\nabla\rho),$$

$$\Delta\rho + (a - 1) = 0, \tag{6.2.5}$$

where $\chi > 0$ and the other coefficients have been rescaled.

For the system (6.2.5) subject to the boundary data (6.2.3), Jäger & Luckhaus (1992) show that when χ is small there exists a unique smooth global positive solution a when the initial data is smooth. When χ is large enough and the spatial dimension is greater than or equal to two, there are solutions a which blow-up in a finite time. The function ρ may be calculated from $(6.2.5)_2$ once a knowledge of a is acquired. In fact, Jäger & Luckhaus (1992) prove the following theorem:

Theorem. (Jäger & Luckhaus (1992)).
(a) There exists a critical number $c(\Omega)$ such that $\alpha\chi\bar{a}_0 < c(\Omega)$ implies that there exists a unique, smooth positive solution to (6.2.5) for all time.
(b) Let Ω be a disc. There exists a positive number c^* such that if $\alpha\chi\bar{a}_0 > c^*$ then radially symmetric positive data may be selected in order that the solution a blows up in a finite time in the centre of the disc.

The blow-up result is interesting in that the explosive behaviour is due to the chemotaxis term which effectively acts as a force on a diffusion equation, see $(6.2.5)_1$.

The proof of part (b) of the above theorem given by Jäger & Luckhaus (1992) uses a comparison argument. They deal with radially symmetric solutions and define the function U by

$$U(r,t) = \int_0^{\sqrt{r}} [a(\phi,t) - 1]\,\phi\,d\phi, \qquad (6.2.6)$$

where ϕ denotes the radial variable $\phi = \sqrt{x_i x_i}$. They show that U satisfies the partial differential equation

$$\frac{\partial U}{\partial t} = 4r\frac{\partial^2 U}{\partial r^2} + \chi\frac{\partial U^2}{\partial r} + \chi U. \qquad (6.2.7)$$

This solution satisfies the boundary and initial conditions

$$U(0,t) = U(R,t) = 0, \qquad (6.2.8)$$

$$U(r,0) = \int_0^{\sqrt{r}} (a_0 - 1)\phi\,d\phi. \qquad (6.2.9)$$

Jäger & Luckhaus (1992) construct a comparison (sub)solution W which satisfies the inequality

$$\frac{\partial W}{\partial t} \le 4r\frac{\partial^2 W}{\partial r^2} + \chi\frac{\partial W^2}{\partial r} + \chi W, \qquad (6.2.10)$$

together with the boundary conditions

$$W(0,t) = W(R,t) = 0,$$

and the initial data restriction

$$W(r,0) \le U(r,0).$$

The function W they construct has form

$$W(r,t) = \frac{\lambda r}{r + \tau^3}, \qquad \text{for } r < r_1,$$

$$W(r,t) = \zeta\left[1 - r - \frac{(r_2 - r)_+^2}{r_2}\right], \qquad \text{for } r_1 \le r,$$

where $(\cdot)_+$ means the positive part, $\tau = r_0 - bt$,

$$\zeta = \frac{\lambda r_1}{(r_1 + \tau^3)\left[1 - r_1 - (r_2 - r_1)^2/r_2\right]},$$

and the numbers r_1, r_2 are such that $0 < r_1 < r_2 < 1$. The terms λ, b and r_0 are chosen in order that W satisfies inequality (6.2.10).

Jäger & Luckhaus (1992) show this function W is such that

$$\limsup_{\substack{t \to \bar{t} \\ r < \epsilon}} W(r,t) \ge \omega > 0,$$

for each $\epsilon > 0$. Whence, blow-up occurs at some time $t^* \le \bar{t}$, and at the spatial point $r = 0$.

6.3 Blow-Up in Chemotaxis with a Single Diffusion Term

Levine & Sleeman (1995) present an interesting study for a system arising effectively out of the Keller & Segel system of partial differential equations (6.1.6). In fact, Levine & Sleeman (1995) refer to a preprint by Othmer & Stevens:
Othmer, H.G. & Stevens, A. Aggregation, blow-up and collapse: the ABCs of taxis and reinforced walks.

I have not seen this preprint and what I write is based on Levine & Sleeman's (1995) article.

According to Levine & Sleeman (1995), Othmer and Stevens are interested in explaining, *"How are the microscopic details of detection of cells to stimuli and their response reflected in the macroscopic parameters of a continuous description?"* Also, they wish to explain, *"Is aggregation possible without long range signalling via a diffusible attractant?"*

Othmer & Stevens develop several mathematical models to begin to provide an explanation of these questions and analyse their models numerically. The paper of Levine & Sleeman (1995) derives rigorous analysis results for a system of Othmer & Stevens and their results confirm the computational deductions deduced by Othmer & Stevens.

To be precise, the article of Levine & Sleeman (1995) investigates the following Othmer & Stevens system, which can be seen to have a similarity with the partial differential equation system (6.1.6),

$$\frac{\partial a}{\partial t} = D \frac{\partial}{\partial x}\left(a \frac{\partial}{\partial x}\left[\log \frac{a}{\Phi(\rho)}\right]\right),$$
$$\frac{\partial \rho}{\partial t} = R(a, \rho), \tag{6.3.1}$$

on the one-dimensional spatial region $(0, \ell)$, with $t > 0$, where the functions Φ and R are given by

$$\Phi(\rho) = \left(\frac{\rho + \beta}{\rho + \gamma}\right)^{\xi}, \tag{6.3.2}$$

$$R(a, \rho) = \lambda \frac{a\rho}{k_1 + \rho} + \frac{\gamma_r a}{k_2 + a} - \mu\rho. \tag{6.3.3}$$

The terms $\beta, \gamma, k_1, k_2, \gamma_r$ and μ are all non-negative constants, D and λ are positive constants, and $\xi \neq 0$. The boundary conditions satisfied by a and ρ are that

$$\frac{\partial}{\partial x}\left[\log \frac{a}{\Phi(\rho)}\right] = 0, \qquad \text{for } x = 0, \ell. \tag{6.3.4}$$

The initial data is given as

$$a(x,0) = a_0(x), \qquad \rho(x,0) = \rho_0(x), \qquad x \in [0, \ell]. \qquad (6.3.5)$$

We remark again that the boundary initial value problem (6.3.1)–(6.3.5) is derived in the preprint of H.G. Othmer and A. Stevens.

Equation (6.3.1) may be written as

$$\frac{\partial a}{\partial t} = D\frac{\partial^2 a}{\partial x^2} - \xi D(\gamma - \beta)\frac{\partial}{\partial x}\left[\frac{a}{(\rho + \gamma)(\rho + \beta)}\frac{\partial \rho}{\partial x}\right], \qquad (6.3.6)$$

and in this form the similarity with equation $(6.1.6)_1$ may be observed. Notice that there is no diffusion in equation $(6.3.1)_2$.

Levine & Sleeman (1995) note that equation $(6.3.1)_1$ is parabolic in a and so $a(x, t) \geq 0$. We shall (essentially follow Levine & Sleeman (1995) and) refer to (6.3.1)–(6.3.5) as the OS system. In fact, Levine & Sleeman (1995) investigate simplified versions of the OS system. They argue that when the coefficients γ and β together with the function ρ are such that $\gamma \gg \rho \gg \beta$, or $\beta \gg \rho \gg \gamma$, one may take $\Phi = 1/\rho^\xi$, then for $\gamma_r = 0$ and $R(a, \rho) = \lambda a\rho - \mu\rho$, the OS system becomes

$$\frac{\partial a}{\partial t} = D\frac{\partial^2 a}{\partial x^2} + \xi D\frac{\partial}{\partial x}\left(\frac{a}{\rho}\frac{\partial \rho}{\partial x}\right),$$

$$\frac{\partial \rho}{\partial t} = \lambda a\rho - \mu\rho, \qquad (6.3.7)$$

with the boundary conditions

$$\frac{\xi}{\rho}\frac{\partial \rho}{\partial x} + \frac{1}{a}\frac{\partial a}{\partial x} = 0, \qquad \text{for } x = 0, \ell, \qquad (6.3.8)$$

and the initial conditions

$$a(x,0) = a_0(x) > 0, \qquad \rho(x,0) = \rho_0(x) > 0, \qquad x \in [0, \ell]. \qquad (6.3.9)$$

Another OS system studied by Levine & Sleeman (1995) examines the situation when $\rho \gg \gamma \gg \beta$ and then the boundary initial value problem becomes

$$\frac{\partial a}{\partial t} = D\frac{\partial^2 a}{\partial x^2} + \xi D\frac{\partial}{\partial x}\left(\frac{a}{\rho^2}\frac{\partial \rho}{\partial x}\right), \qquad x \in (0, \pi), \, t > 0,$$

$$\frac{\partial \rho}{\partial t} = \lambda a\rho - \mu\rho, \qquad x \in (0, \pi), \, t > 0,$$

$$\frac{\xi}{\rho^2}\frac{\partial \rho}{\partial x} + \frac{1}{a}\frac{\partial a}{\partial x} = 0, \qquad x = 0, \pi, \, t > 0, \qquad (6.3.10)$$

$$a(x,0) = a_0(x) > 0, \qquad \rho(x,0) = \rho_0(x) > 0, \qquad x \in [0, \pi].$$

To make progress with the boundary initial value problem (6.3.7)–(6.3.9), Levine & Sleeman (1995) rescale x and t in such a way that $D = 1$ and $\ell = \pi$. They then transform a and ρ into the variables

$$\tilde{a} = \lambda a, \qquad \tilde{\rho} = e^{\mu t}\rho,$$

so that the partial differential equations (6.3.7) become

$$\frac{\partial \tilde{a}}{\partial t} = \frac{\partial^2 \tilde{a}}{\partial x^2} + \xi \frac{\partial}{\partial x}\left(\tilde{a}\Big(\frac{\partial}{\partial x}\log \tilde{\rho}\Big)\right),$$

$$\frac{\partial \tilde{\rho}}{\partial t} = \tilde{a}\tilde{\rho}. \tag{6.3.11}$$

They then introduce the function $\psi(x,t)$ by

$$\psi(x,t) = \log \tilde{\rho}(x,t).$$

Defining the differential operator $\mathcal{L}$ by

$$\mathcal{L} \equiv \frac{\partial^2 \psi}{\partial t^2} - \xi \frac{\partial}{\partial x}\left(\frac{\partial \psi}{\partial x}\frac{\partial \psi}{\partial t}\right),$$

this leads to the following boundary initial value problem for ψ,

$$\mathcal{L}\psi = \frac{\partial^3 \psi}{\partial t \partial x^2}, \qquad x \in (0,\pi),\ t > 0,$$

$$\xi\frac{\partial \psi}{\partial x}\frac{\partial \psi}{\partial t} + \frac{\partial^2 \psi}{\partial x \partial t} = 0, \qquad x = 0,\pi,\ t > 0, \tag{6.3.12}$$

$$\psi(x,0) = \psi_0(x), \qquad \frac{\partial \psi}{\partial t}(x,0) = \tilde{a}_0(x) = \psi_1(x), \qquad x \in (0,\pi),$$

where ψ_0 and ψ_1 are known.

Levine & Sleeman (1995) note that $\mathcal{L}$ is a hyperbolic operator at the point (x,t) when

$$\left(\frac{\partial \psi}{\partial x}\right)^2 + 4\xi\frac{\partial \psi}{\partial t} > 0. \tag{6.3.13}$$

They define $\mathcal{L}$ to be elliptic at (x,t) when the sign is reversed in (6.3.13). Their study of solution behaviour in the hodograph plane, i.e. the (ψ_x, ψ_t) plane is revealing.

We here concentrate on the blow-up results of Levine & Sleeman (1995). We remark, however, that their paper is much wider than this and considers also collapse of solutions, how the solution displays aggregation, and how shock waves form. In particular, they include an account of a "change of type" instability, where the operator changes from hyperbolic to elliptic in the hodograph plane. This instability could have been described in the next chapter where we concentrate on instabilities due to change of type in partial differential equations. It is a very interesting instability, but we refer to the paper of Levine & Sleeman (1995) for details of this.

To return to the blow-up question for the boundary initial value problem (6.3.12) we note that motivated by initial data given in Othmer & Stevens, Levine & Sleeman (1995) choose $\xi = -1$ and then transform ψ to u, where u is defined by the relation $\psi = t + u$. The case $\xi = -1$ is a degenerate case in (6.3.13) which Levine & Sleeman (1995) refer to as a parabolic line of degeneracy. The partial differential equations and boundary conditions the function u must satisfy are

$$\frac{\partial^2 u}{\partial t^2} + \frac{\partial^2 u}{\partial x^2} - \frac{\partial^3 u}{\partial t \partial x^2} = -\frac{\partial}{\partial x}\left(\frac{\partial u}{\partial x}\frac{\partial u}{\partial t}\right), \qquad x \in (0, \pi), \ t > 0, \quad (6.3.14)$$

$$\frac{\partial u}{\partial x}\left(\frac{\partial u}{\partial t} + 1\right) = \frac{\partial^2 u}{\partial t \partial x}, \qquad x = 0, \pi. \qquad (6.3.15)$$

At this point Levine & Sleeman (1995) produce what is a remarkable solution. Even though the partial differential equation (6.3.14) is non-linear, they seek a separation of variables solution of form

$$u(x, t) = \sum_{n=1}^{\infty} a_n \exp\left(cNnt\right)\cos\left(Nnx\right), \qquad (6.3.16)$$

where N is a fixed positive integer and c is a real number. They observe that this expression for u satisfies the boundary condition (6.3.15) and then calculate the left and right hand sides of (6.3.14) to find

$$\frac{\partial^2 u}{\partial t^2} + \frac{\partial^2 u}{\partial x^2} - \frac{\partial^3 u}{\partial t \partial x^2}$$
$$= N^2 \sum_{n=1}^{\infty} a_n n^2 (c^2 + Nnc - 1) \exp\left(cNnt\right)\cos\left(Nnx\right), \qquad (6.3.17)$$

and

$$-\frac{\partial}{\partial x}\left(\frac{\partial u}{\partial x}\frac{\partial u}{\partial t}\right) = \frac{1}{2}N^3 c \sum_{n=2}^{\infty} n\left[\sum_{k=1}^{n-1} k(n-k)a_k a_{n-k}\right] \exp\left(cNnt\right)\cos\left(Nnx\right). \qquad (6.3.18)$$

They equate (6.3.17) and (6.3.18) and deduce that for $n = 1$

$$a_1(c^2 + Nc - 1) = 0,$$

and for $n \geq 2$

$$n(c^2 + Nc - 1)a_n = \frac{1}{2}Nc \sum_{k=1}^{n-1} k(n-k)a_k a_{n-k}.$$

This leads Levine & Sleeman (1995) to produce the *exact solution* $\psi(x, t)$ to (6.3.12), for $\xi = -1$, of form

$$\psi(x, t) = t + 2 \sum_{n=1}^{\infty} \frac{\epsilon^n}{n} \exp\left(cNnt\right)\cos\left(Nnx\right), \qquad (6.3.19)$$

where ϵ is arbitrary. They observe that the series converges absolutely and uniformly on compact sets of $[0, \pi] \times (0, T)$ provided

$$t < T(\epsilon, N) = -\frac{\log |\epsilon|}{Nc}. \tag{6.3.20}$$

The function a may then be shown to have form

$$a(x, t) = 1 + 2Nc \sum_{n=1}^{\infty} \epsilon^n \exp(cNnt) \cos(Nnx), \tag{6.3.21}$$

which is a valid solution for t restricted by (6.3.20).

Levine & Sleeman (1995) also note that (6.3.21) can be written in terms of the complex variable

$$z = Nct + \log \epsilon + iNx$$

as

$$a(x, t) = 1 + Nc[w(z) + w(\bar{z})],$$

where

$$w(z) = \frac{e^z}{1 - e^z}.$$

This gives the relations

$$\psi(x, t) = t - \log\left[1 - 2\epsilon e^{Nct} \cos(Nx) + \epsilon^2 e^{2Nct}\right], \tag{6.3.22}$$

$$a(x, t) = 1 - 2Nc\epsilon\, e^{Nct} \left[\frac{\epsilon\, e^{Nct} - \cos(Nx)}{1 - 2\epsilon\, e^{Nct} \cos(Nx) + \epsilon^2 e^{2Nct}}\right]. \tag{6.3.23}$$

Levine & Sleeman (1995) investigate this solution in detail and deduce that blow-up occurs in finite time at

$$T = -\frac{\log |\epsilon|}{Nc}$$

and at the point $\bar{x}$ satisfying

$$\epsilon\, e^{NcT} \cos(N\bar{x}) = 1.$$

If $N = 2$ and $\epsilon < 0$ they observe that blow-up occurs at the single point $(\pi/2, T)$. These writers also note that blow-up occurs on the "parabolic" boundary where

$$\left(\frac{\partial \psi}{\partial x}\right)^2 - 4\frac{\partial \psi}{\partial t} = 0.$$

Levine & Sleeman (1995) give an explanation of the precise reason for the blow-up being due to the non-linear term $(u_x u_t)_x$ in equation (6.3.14).

To conclude this chapter we remark that the paper of Levine & Sleeman (1995) is long and contains many more results than are described here. In particular, they consider instability due to the equation changing from being hyperbolic to being elliptic, a topic we investigate in other contexts in the next chapter.

In the remaining two chapters we change direction a little and do not study partial differential equations in which blow-up in finite time has been observed. We do, however, include an exposition of very important recent studies in which very rapid instabilities have been found. The problems we report on are all in mechanics and involve partial differential equations in a continuum mechanics setting.

7. Change of Type

7.1 Instability in a Hypoplastic Material

In this chapter we give an exposition of three physically interesting recent pieces of work which deal with very rapid instabilities due to the nature of the partial differential equations changing from being hyperbolic to being elliptic.

There are many examples of occurrences in mechanics of instabilities due to a change of type in the governing partial differential equation or system of partial differential equations. Joseph & Saut (1990) is an interesting review containing several such applied situations. The standard example which is often used to illustrate this type of instability is to consider the Cauchy problem for the Laplace equation. Hadamard (1922) constructed this example to show how the Cauchy problem for the Laplace equation possesses an unstable solution, see e.g. Ames & Straughan (1997), pp. 230–231, Joseph & Saut (1990), p. 193. The example takes Laplace's equation

$$\frac{\partial^2 v}{\partial t^2} + \frac{\partial^2 v}{\partial x^2} = 0, \tag{7.1.1}$$

on the domain $(x,t) \in \mathbf{R} \times (0, \infty)$, with the initial (Cauchy) data

$$v(x,0) = 0, \qquad \frac{\partial v}{\partial t}(x,0) = g(x), \tag{7.1.2}$$

where $g(x)$ is the function

$$g(x) = \frac{1}{\xi^p} \sin \xi x, \tag{7.1.3}$$

for some $p > 0$. Note that the data function $g(x)$ is bounded for all $x \in \mathbf{R}$ and $g \to 0$ as $\xi \to \infty$. Thus, $g \to 0$ for vanishingly small disturbances of wavelength $2\pi/\xi$, $\xi \to \infty$. The solution to (7.1.1)–(7.1.3) is

$$v(x,t) = \frac{1}{\xi^{1+p}} \sin \xi x \sinh \xi t. \tag{7.1.4}$$

This solution is unbounded as $t \to \infty$, no matter how small g is. Also, even for $t > 0$ but small, the solution can still be arbitrarily large for small wavelengths

$2\pi/\xi$, i.e. as $\xi \to \infty$. Joseph & Saut (1990) call this type of instability, which makes the problem (7.1.1)–(7.1.3) improperly posed, a *Hadamard instability*.

From a numerical point of view, such instabilities are very serious, as Bellomo & Preziosi (1995), p. 351, point out. By making the discretization finer one can still pick up the small wavelength disturbances and these lead to serious numerical instability problems. Thus, in any boundary initial value problem where the partial differential equation changes from being hyperbolic to being elliptic one has the possibility of instabilities in the elliptic zone of Hadamard type.

Joseph & Saut (1990) use the following example to illustrate such behaviour in a mechanics environment. Consider the partial differential equations

$$\frac{\partial u}{\partial t} = \frac{\partial}{\partial x}\sigma(v),$$
$$\frac{\partial v}{\partial t} = \frac{\partial u}{\partial x},$$

(7.1.5)

where the function $\sigma(v)$ has an inverted S shape, cf. Fig. 7.1.

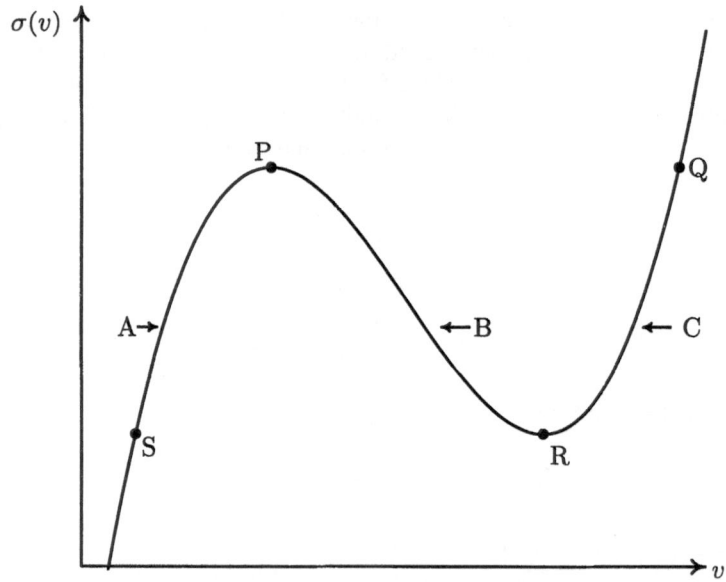

Fig. 7.1 The function $\sigma(v)$.

Such a function $\sigma(v)$ is common place in phase change theories, and for example in a van der Waals gas.

From (7.1.5) one can eliminate u and derive the following non-linear partial differential equation for v,

$$\frac{\partial^2 v}{\partial t^2} - \sigma'(v)\frac{\partial^2 v}{\partial x^2} = \sigma''(v)\left(\frac{\partial v}{\partial x}\right)^2 . \tag{7.1.6}$$

If the function σ has the behaviour shown in Fig. 7.1, then on the branches A or C, $\sigma' > 0$ and equation (7.1.6) is hyperbolic. If, however, one is on branch B then $\sigma' < 0$ and so equation (7.1.6) is elliptic. Therefore, when v is in the B range the solution to (7.1.5) is subject to catastrophic Hadamard instabilities. The function $\sigma(v)$ allows for hysteresis. When σ increases with v on branch A the function can jump from P to Q. As v is allowed to decrease on branch C the function σ can decrease down to R at which point it may jump to S.

These ideas are amplified in Joseph & Saut (1990) and ways of regularizing the equation(s) to overcome the instability are discussed. Joseph & Saut (1990) discuss applications in phase change models, and also in viscoelastic flows. Joseph et al. (1985) describe many change of type instabilities in a variety of situations involving viscoelastic fluids.

Another interesting example of change of type instability occurs in chemotaxis, as discussed by Levine & Sleeman (1995) using the model discussed in section 6.3. Schaeffer (1990, 1992), Schaeffer & Shearer (1997), Shearer & Schaeffer (1994), and Lianjun An & Schaeffer (1992) give detailed accounts of instabilities in granular materials caused by change of type, and other effects. Particular attention to elastic plastic behaviour is given by Lianjun An (1994), and Lianjun An & Peirce (1994, 1995) study instabilities in a theory for an elastic plastic material described by using a Cosserat theory, and by accounting for the presence of microscale voids.

The effect of Hadamard instabilities is of issue also in stochastic differential equations and in inverse problems. Bellomo & Preziosi (1995), pp. 377–415, discuss several interesting problems of this nature in population dynamics and in an inverse problem for a non-linear theory of diffusion. They also include details of numerical routines for the solution of such problems. Further details of this are given in the books of Bellomo & Riganti (1987) and Bellomo et al. (1992), and in the articles of Preziosi (1993), Preziosi & de Socio (1991), and Preziosi et al. (1992).

In this subsection we include an account of work of Osinov & Gudehus (1996) who discover a change of type instability in a theory for a hypoplastic material. The application of hypoplastic materials to slip of saturated sands and soils has been considered in the engineering literature, and several references to this are given in Osinov & Gudehus (1996). Other approaches to landslides are, however, discussed in Hutter et al. (1987), Vulliet & Hutter (1988a, 1988b), Samtani et al. (1994), Desai et al. (1995), and Vulliet (1995), where many other references to work in this area may be found.

Osinov & Gudehus (1996) develop a theory for a fully saturated body of granular material. They suppose this material consists of a solid fraction, of volume V_s, and a liquid fraction, of volume V_ℓ, but they assume a common velocity and deal with a total stress. Their theory is based on the idea of an incompressible material and they adopt the following equations for the description of the material

$$\rho\left(\frac{\partial \mathbf{v}}{\partial t} + (\mathbf{v} \cdot \nabla)\mathbf{v}\right) = \rho\mathbf{f} + \text{div } \mathbf{T}_{\text{total}},$$

$$\text{div } \mathbf{v} = 0,$$

(7.1.7)

where $\mathbf{v}$ is the velocity, ρ the density, $\mathbf{f}$ the body force, and $\mathbf{T}_{\text{total}}$ denotes the total Cauchy stress tensor. The theory of hypoplasticity they develop uses the stress tensor for the solid constituent, $\mathbf{T}$. The constitutive theory for this satisfies the following equation,

$$\overset{\circ}{\mathbf{T}} = \mathbf{H}(\mathbf{T}, \mathbf{D}, e),$$

(7.1.8)

where e is the void ratio $e = V_\ell/V_s$ and $\mathbf{D}$ is the symmetric part of the velocity gradient,

$$D_{ij} = \frac{1}{2}\left(\frac{\partial v_i}{\partial x_j} + \frac{\partial v_j}{\partial x_i}\right).$$

The notation $\overset{\circ}{\mathbf{T}}$ stands for the Jaumann derivative of $\mathbf{T}$ defined by

$$\overset{\circ}{\mathbf{T}} = \frac{d\mathbf{T}}{dt} + \mathbf{T}\mathbf{W} - \mathbf{W}\mathbf{T},$$

(7.1.9)

where $d\mathbf{T}/dt$ is the material derivative

$$\frac{d\mathbf{T}}{dt} = \frac{\partial \mathbf{T}}{\partial t} + (\mathbf{v} \cdot \nabla)\mathbf{T},$$

and $\mathbf{W}$ is the skew-symmetric part of the velocity gradient

$$W_{ij} = \frac{1}{2}\left(\frac{\partial v_i}{\partial x_j} - \frac{\partial v_j}{\partial x_i}\right).$$

The constitutive function $\mathbf{H}$ is developed in Osinov & Gudehus (1996), although some details are presented below.

The liquid pressure and the grain stress are written as a static part $p_\ell^0(\mathbf{x}), \mathbf{T}^0(\mathbf{x})$ plus a dynamic part $p_\ell(\mathbf{x}, t), \mathbf{T}(\mathbf{x}, t)$. The static part satisfies the equilibrium equation

$$\text{div } \mathbf{T}^0 - \nabla p_\ell^0 + \rho\mathbf{f} = 0,$$

(7.1.10)

while the dynamic part satisfies the equation

$$\rho\left(\frac{\partial v_i}{\partial t} + v_j\frac{\partial v_i}{\partial x_j}\right) = -\frac{\partial p_\ell}{\partial x_i} + \frac{\partial}{\partial x_j}T_{ji}.$$

(7.1.11)

Osinov & Gudehus (1996) consider disturbances of plane shear wave form and so examine solutions of the form

$$v_1 = v_3 = 0, \quad v_2 = v_2(x_1, t),$$

or, if we use the notation $\mathbf{x} = (x, y, z)$ and $\mathbf{v} = (u, v, w)$, they assume

$$\mathbf{v} = (0, v(x, t), 0). \tag{7.1.12}$$

Such a representation does satisfy the incompressibility condition $(7.1.7)_2$. Osinov & Gudehus (1996) assume $T_{13}^0, T_{23}^0, T_{13}$ and T_{23} are zero, while the other components of T_{ij}^0, T_{ij} and $p = p_\ell$ depend only on x and t. Then the equations (7.1.7) and (7.1.8) yield the following six partial differential equations of Osinov & Gudehus (1996) theory,

$$
\begin{aligned}
0 &= -\frac{\partial p}{\partial x} + \frac{\partial T_{11}}{\partial x}, \\
\rho \frac{\partial v}{\partial t} &= \frac{\partial T_{12}}{\partial x}, \\
\frac{\partial T_{11}}{\partial t} &= -(T_{12}^0 + T_{12})\frac{\partial v}{\partial x} + H_{11}\left(\mathbf{T}^0 + \mathbf{T}, \frac{\partial v}{\partial x}, e\right), \\
\frac{\partial T_{12}}{\partial t} &= \frac{1}{2}(T_{11}^0 + T_{11} - T_{22}^0 - T_{22})\frac{\partial v}{\partial x} + H_{12}\left(\mathbf{T}^0 + \mathbf{T}, \frac{\partial v}{\partial x}, e\right), \\
\frac{\partial T_{22}}{\partial t} &= (T_{12}^0 + T_{12})\frac{\partial v}{\partial x} + H_{22}\left(\mathbf{T}^0 + \mathbf{T}, \frac{\partial v}{\partial x}, e\right), \\
\frac{\partial T_{33}}{\partial t} &= H_{33}\left(\mathbf{T}^0 + \mathbf{T}, \frac{\partial v}{\partial x}, e\right).
\end{aligned}
\tag{7.1.13}
$$

The system of equations (7.1.13) is a system of six partial differential equations in the variables $p, v, T_{11}, T_{12}, T_{22}$ and T_{33} which governs the behaviour of a plane shear flow in an Osinov & Gudehus (1996) hypoplastic material.

Osinov & Gudehus (1996) assume that the dynamic stress variation is small in comparison with the size of the stress in the static state so that they may use the approximation

$$\mathbf{H}\left(\mathbf{T}^0 + \mathbf{T}, \frac{\partial v}{\partial x}, e\right) \approx \mathbf{H}\left(\mathbf{T}^0, \frac{\partial v}{\partial x}, c\right). \tag{7.1.14}$$

For the constitutive function $\mathbf{H}$ in (7.1.8) they take

$$\mathbf{H}(\mathbf{T}, \mathbf{D}, e) = f_s\left[\frac{1}{a^2}\mathbf{D} + \mathrm{tr}(\hat{\mathbf{T}}\mathbf{D})\hat{\mathbf{T}} + \frac{1}{a}f_d(\hat{\mathbf{T}} + \hat{\mathbf{T}}^*)\|\mathbf{D}\|\right], \tag{7.1.15}$$

where the functions f_s, f_d and a may depend on e and the invariants of T_{ij}. The other terms in (7.1.15) are

$$\hat{\mathbf{T}} = \frac{\mathbf{T}}{\mathrm{tr}\,\mathbf{T}}, \qquad \hat{\mathbf{T}}^* = \hat{\mathbf{T}} - \frac{1}{3}\mathbf{I}, \qquad \|\mathbf{D}\| = \sqrt{\mathrm{tr}\,(\mathbf{DD})}.$$

The hypoplastic material so defined does not satisfy a linear law. By using further physical arguments appropriate to granular media Osinov & Gudehus (1996) show that equation (7.1.15) used in the partial differential equations $(7.1.13)_{3-6}$ may be conveniently rewritten with the notation

$$\left|\frac{\partial v}{\partial x}\right| = \frac{\partial v}{\partial x} \operatorname{sgn}\left(\frac{\partial v}{\partial x}\right).$$

The system of partial differential equations is then reduced to

$$0 = -\frac{\partial p}{\partial x} + \frac{\partial T_{11}}{\partial x},$$

$$\rho \frac{\partial v}{\partial t} = \frac{\partial T_{12}}{\partial x},$$

$$\frac{\partial T_{11}}{\partial t} = \kappa_1 \frac{\partial v}{\partial x} + \kappa_2 \left|\frac{\partial v}{\partial x}\right|,$$

$$\frac{\partial T_{12}}{\partial t} = \kappa_3 \frac{\partial v}{\partial x} + \kappa_4 \left|\frac{\partial v}{\partial x}\right|, \qquad (7.1.16)$$

$$\frac{\partial T_{22}}{\partial t} = \kappa_5 \frac{\partial v}{\partial x} + \kappa_6 \left|\frac{\partial v}{\partial x}\right|,$$

$$\frac{\partial T_{33}}{\partial t} = \kappa_7 \frac{\partial v}{\partial x} + \kappa_8 \left|\frac{\partial v}{\partial x}\right|,$$

where the coefficients $\kappa_1, \ldots, \kappa_8$ depend on the values of the stress components $T_{11}^0, T_{12}^0, T_{22}^0$ and T_{33}^0 in the static state.

Osinov & Gudehus (1996) observe that if the initial state is oriented in such a way that the principal axes of T_{ij}^0 coincide with the coordinate axes (x, y, z) then the coefficients $\kappa_1, \kappa_4, \kappa_5$ and κ_7 in (7.1.16) vanish. They then investigate the coupled system of partial differential equations which separates out from (7.1.16), namely

$$\rho \frac{\partial v}{\partial t} = \frac{\partial T_{12}}{\partial x},$$

$$\frac{\partial T_{12}}{\partial t} = \kappa_3 \frac{\partial v}{\partial x}. \qquad (7.1.17)$$

Extensive details of the properties of a wave which propagates in a principal stress direction according to (7.1.17) are calculated by Osinov & Gudehus (1996).

Of interest in this monograph is the further analysis of Osinov & Gudehus (1996) which does not take $T_{12}^0 = 0$. They use the two equations $(7.1.16)_2$ and $(7.1.16)_4$ which are non-linear because of the absolute value in $(7.1.16)_4$. These equations may be rewritten as the system of partial differential equations

$$\rho\frac{\partial v}{\partial t} = \frac{\partial T_{12}}{\partial x},$$

$$\frac{\partial T_{12}}{\partial t} = (\kappa_3 + \kappa_4)\frac{\partial v}{\partial x}, \qquad \text{when} \quad \frac{\partial v}{\partial x} > 0, \qquad (7.1.18)$$

$$\frac{\partial T_{12}}{\partial t} = (\kappa_3 - \kappa_4)\frac{\partial v}{\partial x}, \qquad \text{when} \quad \frac{\partial v}{\partial x} < 0.$$

The wavespeed, c, of a wave which results from these equations is not single valued due to the need to have knowledge of $\text{sgn}\,(\partial v/\partial x)$. Instead, the wave propagation must be described by the two wavespeeds c_1 and c_2 given by

$$c_1 = \sqrt{\frac{\kappa_3 + \kappa_4}{\rho}}, \qquad \text{when} \quad \frac{\partial v}{\partial x} > 0,$$

$$\qquad\qquad\qquad\qquad\qquad\qquad\qquad\qquad (7.1.19)$$

$$c_2 = \sqrt{\frac{\kappa_3 - \kappa_4}{\rho}}, \qquad \text{when} \quad \frac{\partial v}{\partial x} < 0.$$

The coefficients κ_3 and κ_4 are

$$\kappa_3 = \frac{1}{2}(T_{11}^0 - T_{22}^0) + \frac{1}{2}h_1 + h_2\big(\hat{T}_{12}^0\big)^2, \qquad \kappa_4 = \sqrt{2}\,h_3\hat{T}_{12}^0, \qquad (7.1.20)$$

where h_1, h_2, h_3 define the specific form of $\mathbf{H}$ used. A non-zero initial stress thus leads to a complicated wave with a *split* wavespeed depending on the velocity gradient. Osinov & Gudehus (1996) develop this idea in detail for a specific sand from Karlsruhe.

Due to the forms (7.1.19) this allows the possibility of the wavespeed becoming imaginary which is equivalent to the system (7.1.18) changing from being hyperbolic to being elliptic and hence giving rise to instability. As Osinov & Gudehus (1996) remark "*a small boundary disturbance can result in a flow of the whole body which cannot be prevented by boundary control. Such a mechanism of the loss of stability can be regarded as a model of spontaneous liquefaction.*"

While Osinov & Gudehus (1996) develop their condition for instability on the basis of a wave analysis, as briefly outlined above, they also use a Hill (1958) type analysis on their equations. They observe that this does not deal directly with the dynamic stability concept. However, they interestingly show that the criterion they derive with the aid of a Hill (1958) type of analysis agrees with what they find using wave theory.

I believe that the hypoplasticity theory used in Osinov & Gudehus (1996) and the dynamic analysis they develop is very interesting and ought to prove useful in the field of saturated sand/soil slip dynamics.

Osinov (1997, 1998) develops a theory for hypoplasticity in a granular body which is not fully saturated with fluid. He carefully investigates the well-posedness of the theory and how and when waves can propagate.

7.2 Instability in a Viscous Plastic Model for Sea Ice Dynamics

We now discuss another instance where the change of type instability is important in a further area of mechanics which can play an important role in life.

Gray & Killworth (1995) encounter the phenomenon of instability due to a hyperbolic equation becoming elliptic in their study of sea ice dynamics. They write that the sea ice is composed of thin horizontal layers separated by water. This sea ice pack floats on top of the ocean. Stresses in the sea ice pack can lead to their model yielding an elliptic equation which gives instability to short wavelengths and so numerically causes serious problems in a computational simulation. However, when the flow velocity in the ice pack is convergent the ice strength reduces and their model is *non-linearly* stable.

The equations used by Gray & Killworth (1995) to describe a sea ice pack consist of partial differential equations to describe the evolutionary behaviour of the velocity field, $\mathbf{v}$, in the pack, and quantities h^i and A which are, respectively, a measure of the ice volume in a given domain and a scaling factor relating h^i to the ice thickness h by the equation $h^i = Ah$. The theory is developed for a spatial region which is in $\mathbf{R}^2$, so A, h^i and v_1, v_2 are functions of x, y and t. A reaction–diffusion system is written for A and h^i, with a momentum balance postulated involving the rate of change of v_α, $\alpha = 1, 2$. While the theory is developed for two-dimensional behaviour of an ice sheet of thickness h, Gray & Killworth (1995) direct their attention to a one-dimensional solution of their equations which is independent of the variable y, and they take $v_2 = 0$. It is convenient to set $v_1 = v$. By adopting a visco-plastic constitutive theory for the stress in the sea ice they derive the following system of partial differential equations for the evolution of a one-dimensional disturbance,

$$\frac{\partial A}{\partial t} + v\frac{\partial A}{\partial x} + A\frac{\partial v}{\partial x} = 0, \tag{7.2.1}$$

$$\frac{\partial h^i}{\partial t} + v\frac{\partial h^i}{\partial x} + h^i\frac{\partial v}{\partial x} = 0, \tag{7.2.2}$$

$$\rho^I h^i \left(\frac{\partial v}{\partial t} + v\frac{\partial v}{\partial x}\right) = -\rho^W C_L^W v + \frac{1}{2}\left[\frac{\sqrt{1+e^2}}{e}\,\mathrm{sgn}\,D - 1\right]\frac{\partial p}{\partial x}, \tag{7.2.3}$$

where p is the pressure which is given by the equation

$$p = p^* h^i \exp\left\{-C(1-A)\right\}, \tag{7.2.4}$$

for constants p^* and C. Here ρ^I and ρ^W are ice and water density, C_L^W is a coefficient measuring drag between the sea ice and the ocean, e is the ratio of principal axes of stress, and $D = \partial v/\partial x$.

Gray & Killworth (1995) transform system (7.2.1)–(7.2.3) to a Lagrangian coordinate form in ξ, t, where ξ is the position at $t = 0$ of a particle at x. If we use the same notation for A, h^i and v in the (ξ, t) coordinates then the governing partial differential equations become

$$\frac{\partial A}{\partial t} + \frac{A}{F}\frac{\partial v}{\partial \xi} = 0, \tag{7.2.5}$$

$$\frac{\partial h^i}{\partial t} + \frac{h^i}{F}\frac{\partial v}{\partial \xi} = 0, \tag{7.2.6}$$

and

$$\rho^I h^i \frac{\partial v}{\partial t} = -\rho^W C_L^W v + \frac{1}{2}\left[\frac{\sqrt{1+e^2}}{e}\operatorname{sgn} D - 1\right]\frac{1}{F}\frac{\partial p}{\partial \xi}, \tag{7.2.7}$$

where $F = \partial\chi/\partial\xi(> 0)$ is the deformation gradient for the mapping $x = \chi(\xi, t)$.

Gray & Killworth (1995) analyse the stability of a steady state solution $\bar{h}^i, \bar{A}, \bar{v} = 0$, by writing

$$h^i \to \bar{h}^i + h^i, \qquad A \to \bar{A} + A, \qquad F \to \bar{F} + F, \qquad v \to \bar{v} + v,$$

and from (7.2.5)–(7.2.7) they derive equations for the perturbation quantities h^i, A and v. The perturbation equations are

$$\frac{\partial A}{\partial t} + A\frac{\partial v}{\partial \xi} = 0, \tag{7.2.8}$$

$$\frac{\partial h^i}{\partial t} + h^i\frac{\partial v}{\partial \xi} = 0, \tag{7.2.9}$$

$$\psi_1\frac{\partial v}{\partial t} + \psi_2 v = \psi_3\left[\frac{\partial h^i}{\partial \xi} + C\bar{h}^i\frac{\partial A}{\partial \xi}\right], \tag{7.2.10}$$

where ψ_1, ψ_2 are positive constants while ψ_3 is given by

$$\psi_3 = \frac{1}{2}p^* \exp\left[-C(1 - \bar{A})\right]\left(\frac{\sqrt{1+e^2}}{e}\operatorname{sgn} D - 1\right).$$

For a divergent flow, $D > 0$, and then the coefficient ψ_3 is also positive. However, in convergent flow, $D < 0$, and ψ_3 will be negative. A single equation for v may be found from (7.2.8)–(7.2.10), and this has form

$$\frac{\partial^2 v}{\partial t^2} + \frac{\psi_2}{\psi_1}\frac{\partial v}{\partial t} = -\frac{\psi_3}{\psi_1}\bar{h}^i(1 + C\bar{A})\frac{\partial^2 v}{\partial \xi^2}. \tag{7.2.11}$$

When $\psi_3 < 0$, (7.2.11) is a damped wave equation. This is the case when the velocity field corresponds to a converging flow with $D < 0$. The stress has form, Gray & Killworth (1995)

$$N = \frac{1}{2}p\left(\frac{\sqrt{1+e^2}}{e}\operatorname{sgn} D - 1\right)$$

and thus the stress is in this case compressive. When, however, $\psi_3 > 0$, equation (7.2.11) is elliptic and Hadamard instability is possible. This is the situation for diverging flow, where $D > 0$, in which case the stress is tensile, $N > 0$. Gray & Killworth (1995) actually deduce this by neglecting the damping term in (7.2.11) and working on a fixed domain $\xi \in [0, \pi/2k]$ and using the solution

$$v = v_0 \sin(k\xi) \exp(\omega t).$$

Gray & Killworth (1995) further analyse the possibility of regularizing their equations to allow a numerical solution for a sea ice sheet evolution problem to be acceptable.

7.3 Pressure Dependent Viscosity Flow

Renardy (1986) contains a rigorous study of a solution to the Navier-Stokes equations when the viscosity depends on pressure. He shows that a change of type is possible and instability may occur. As Renardy (1986) and Bair & Khonsari (1996, 1997) point out, fluids with large viscosity dependence on pressure have wide industrial application. In particular, Bair & Khonsari (1997) draw attention to the application of such flows in lubrication theory where an oil is squeezed in a very small gap between metal surfaces.

Bair & Khonsari (1996) have performed experiments on capillary flow employing two different fluids, a polyphenyl ether, and a polyelphaolefin, both of whose viscosity varies enormously with pressure changes. They adopt an exponential relationship for the viscosity variation, of form

$$\mu = \mu_0 e^{\alpha p}, \tag{7.3.1}$$

where μ_0 and α are constants. Their theoretical approach begins with steady flow of an incompressible fluid whose viscosity satisfies (7.3.1). They neglect the non-linear convective terms, but it is convenient to leave them in since their method of analysis still carries through. The equations governing such a flow are

$$\rho u_j \frac{\partial u_i}{\partial x_j} = -\frac{\partial p}{\partial x_i} + \frac{\partial \sigma_{ji}}{\partial x_j},$$
$$\frac{\partial u_i}{\partial x_i} = 0, \tag{7.3.2}$$

where ρ is a constant density, u_i is velocity, p is pressure, and the stress tensor is given by

$$\sigma_{ji} = -p\delta_{ij} + 2\mu d_{ij}. \tag{7.3.3}$$

The quantity d_{ij} is the symmetric part of the velocity gradient, i.e.

$$d_{ij} = \frac{1}{2}\left(\frac{\partial u_i}{\partial x_j} + \frac{\partial u_j}{\partial x_i}\right). \qquad (7.3.4)$$

For steady two-dimensional flow, (7.3.2) become

$$\rho\left(u\frac{\partial u}{\partial x} + v\frac{\partial u}{\partial y}\right) + \frac{\partial p}{\partial x} = 2\frac{\partial}{\partial x}\left(\mu\frac{\partial u}{\partial x}\right) + \frac{\partial}{\partial y}\left[\mu\left(\frac{\partial v}{\partial x} + \frac{\partial u}{\partial y}\right)\right],$$

$$\rho\left(u\frac{\partial v}{\partial x} + v\frac{\partial v}{\partial y}\right) + \frac{\partial p}{\partial y} = 2\frac{\partial}{\partial y}\left(\mu\frac{\partial v}{\partial y}\right) + \frac{\partial}{\partial x}\left[\mu\left(\frac{\partial v}{\partial x} + \frac{\partial u}{\partial y}\right)\right], \qquad (7.3.5)$$

$$\frac{\partial u}{\partial x} + \frac{\partial v}{\partial y} = 0.$$

Equation $(7.3.5)_1$ is expanded using $(7.3.5)_3$ to derive

$$\rho\left(u\frac{\partial u}{\partial x} + v\frac{\partial u}{\partial y}\right) + \frac{\partial p}{\partial x} = \mu\Delta u + 2\frac{\partial \mu}{\partial x}\frac{\partial u}{\partial x} + \frac{\partial \mu}{\partial y}\left(\frac{\partial v}{\partial x} + \frac{\partial u}{\partial y}\right)$$

and then by using the viscosity relation (7.3.1) this leads to

$$\frac{\partial p}{\partial x}\left(1 - 2\alpha\mu\frac{\partial u}{\partial x}\right) = \mu\Delta u - \rho\left(u\frac{\partial u}{\partial x} + v\frac{\partial u}{\partial y}\right) + \alpha\mu\frac{\partial p}{\partial y}\left(\frac{\partial v}{\partial x} + \frac{\partial u}{\partial y}\right). \qquad (7.3.6)$$

A similar calculation starting with equation $(7.3.5)_2$ yields

$$\frac{\partial p}{\partial y}\left(1 - 2\alpha\mu\frac{\partial v}{\partial y}\right) = \mu\Delta v - \rho\left(u\frac{\partial v}{\partial x} + v\frac{\partial v}{\partial y}\right) + \alpha\mu\frac{\partial p}{\partial x}\left(\frac{\partial v}{\partial x} + \frac{\partial u}{\partial y}\right). \qquad (7.3.7)$$

By elimination of p_y between (7.3.6) and (7.3.7), one finds

$$\frac{\partial p}{\partial x} = \frac{A + B}{(1 - 4\alpha^2\mu^2 u_x^2 - \alpha^2\mu^2[v_x + u_y]^2)}, \qquad (7.3.8)$$

where A and B are given by the expressions

$$A = (\mu\Delta u - \rho u u_x - \rho v u_y)(1 - 2\alpha\mu v_y)$$

$$B = \alpha\mu(v_x + u_y)(\mu\Delta v - \rho u v_x - \rho v v_y).$$

A similar expression may be derived for p_y, see Bair & Khonsari (1996).

The above derivation is due to Bair & Khonsari (1996) who then observe that if the numerator in (7.3.8) remains finite it is possible to have singularity formation when the denominator becomes zero. They also relate the vanishing of the denominator to a condition on the shear stress.

Bair & Khonsari (1996) further study two-dimensional flow between parallel planes and then adopt the parabolic velocity profile which holds for viscous flow, i.e. that found in Navier–Stokes theory. This allows them to derive a condition under which the pressure may become infinite.

Bair & Khonsari (1996) apply their analysis to shear band formation which they observe experimentally for flow between parallel planes. Such a phenomenon may have important application in lubrication theory and this is further investigated in Bair & Khonsari (1997).

8. Rapid Energy Growth in Parallel Flows

8.1 Rapid Growth in Incompressible Viscous Flows

In this chapter we report on some recent findings where the kinetic energy for a certain class of fluid flows can have relatively very large transient growth, even though the flow is eventually stable according to linear theory. To facilitate the discussion which follows it is convenient to digress into a very brief account of non-linear energy stability theory in the context of certain hydrodynamic flows.

Under the majority of situations, when an incompressible viscous fluid is heated from below, provided the depth of the fluid layer or the temperature gradient across the layer is large enough, a convective instability in the form of a cellular pattern commences. The theory for this is well documented. For example, accounts of the linearised instability and non-linear energy stability theories for this problem may be found in Straughan (1992). The partial differential equations governing a perturbation in the stability/instability theory for the problem of thermal convection outlined above are

$$\frac{\partial u_i}{\partial t} + u_j \frac{\partial u_i}{\partial x_j} = -\frac{\partial p}{\partial x_i} + \Delta u_i + R\theta k_i,$$

$$\frac{\partial u_i}{\partial x_i} = 0, \tag{8.1.1}$$

$$Pr\left(\frac{\partial \theta}{\partial t} + u_j \frac{\partial \theta}{\partial x_j}\right) = Rw + \Delta\theta,$$

where u_i, θ, p are velocity, temperature, and pressure perturbations, $\mathbf{k} = (0,0,1)$, Pr is the Prandtl number, and $Ra = R^2$ is the Rayleigh number. The partial differential equations (8.1.1) hold on the space–time domain $\mathbf{R}^2 \times \{z \in (0,1)\} \times \{t > 0\}$. The boundary conditions often imposed are

$$u_i = 0, \qquad \theta = 0, \qquad z = 0, 1, \tag{8.1.2}$$

with u_i, θ and p satisfying a plane tiling form in the x and y variables. These boundary conditions correspond to no-slip on the velocity field at the plates $z = 0$ and $z = 1$, and prescribed temperatures at these surfaces. The periodic restriction in the horizontal (x, y) variables reflects the observed cellular patterns which ensue once convection has commenced. The thermal convection scenario we have sketchily outlined is normally called Bénard convection.

For the problem defined by the partial differential equations (8.1.1) and the boundary conditions (8.1.2), the critical Rayleigh number of linear *instability* theory, Ra_L, *coincides exactly* with the critical Rayleigh number of non-linear *stability* theory, Ra_E, see e.g. the account in Straughan (1992). Thus, no subcritical instabilities may arise and the theory of non-linear energy stability demonstrates that the linear instability theory has captured the essential physics of the onset of thermal convection. Thus, the results of both linear instability theory and non-linear energy stability theory are highly useful.

The system of partial differential equations (8.1.1) and boundary conditions (8.1.2) is a special case of the abstract system, defined in a Hilbert space,

$$Au_t + Lu + N(u) + \epsilon Mu = 0. \qquad (8.1.3)$$

Here A is a bounded symmetric linear operator, L is a symmetric linear operator, M is a skew-symmetric linear operator and $N(u)$ is a non-linear operator, which in (8.1.1) is represented by the convective terms $u_j u_{i,j}$ and $u_j \theta_{,j}$. Complete details of this may be found in e.g. Straughan (1992), where the connection between linear and non-linear stability is explored in some depth.

When $\epsilon = 0$ in (8.1.3) then for L a sectorial operator with a compact resolvent, the linear and non-linear stability boundaries still coincide provided $(u, N(u)) \geq 0$ in the inner product associated with the Hilbert space for which (8.1.3) is defined. Details of this may be found in e.g. Straughan (1992), chapter 4. In fact, for M bounded the temporal growth rate of the linear theory for (8.1.3) continues to be real for ϵ small enough and then the energy and linear stability boundaries remain within $O(\epsilon)$ of each other, cf. Davis (1969).

There are, however, many problems in applied mathematics in which ϵ is not small. Good examples of this arise in Bénard convection in a rotating layer, and in the Bénard problem with a superposed magnetic field. The partial differential equations for these problems differ from (8.1.1). For the rotating Bénard problem, $(8.1.1)_{2,3}$ are the same, but $(8.1.1)_1$ is replaced by

$$\frac{\partial u_i}{\partial t} + u_j \frac{\partial u_i}{\partial x_j} = -\frac{\partial p}{\partial x_i} + R\theta k_i + \Delta u_i + T(\mathbf{u} \times \mathbf{k})_i. \qquad (8.1.4)$$

In this equation $\mathbf{k} = (0, 0, 1)$ and T^2 is the Taylor number which measures the magnitude of the angular velocity with which the layer is rotating. The skew-symmetric term $T(\mathbf{u} \times \mathbf{k})$ corresponds to the term ϵMu in equation (8.1.3). The magnetic Bénard problem is governed by the partial differential equations, see e.g. Mulone & Rionero (1993),

$$\frac{\partial u_i}{\partial t} + u_j \frac{\partial u_i}{\partial x_j} = -\frac{\partial p}{\partial x_i} + R\theta k_i + \Delta u_i + P_m h_j \frac{\partial h_i}{\partial x_j} + Q\frac{\partial h_i}{\partial z},$$

$$\frac{\partial u_i}{\partial x_i} = 0,$$

$$P_m \left(\frac{\partial h_i}{\partial t} + u_j \frac{\partial h_i}{\partial x_j} \right) = \Delta h_i + P_m h_j \frac{\partial u_i}{\partial x_j} + Q\frac{\partial u_i}{\partial z}, \qquad (8.1.5)$$

$$\frac{\partial h_i}{\partial x_i} = 0,$$

$$Pr\left(\frac{\partial \theta}{\partial t} + u_i \frac{\partial \theta}{\partial x_i} \right) = Rw + \Delta\theta.$$

The quantity P_m is the magnetic Prandtl number while Q^2 is the Chandrasekhar number, which is a non-dimensional form of the size of the imposed magnetic field. In this case the terms $Q\partial h_i/\partial z$ and $Q\partial u_i/\partial z$ in (8.1.5) correspond to the skew-symmetric piece $\epsilon M u$ in equation (8.1.3).

For the problem of a fluid layer which is heated from below which is simultaneously undergoing a uniform rotation of the layer, the equations governing a perturbation to the steady solution are given by $(8.1.1)_{2,3}$ and (8.1.4). For this situation, when the Taylor number, T, is large, the linear and standard non-linear energy stability results differ greatly, cf. Straughan (1992), Chap. 6. The same is true in the magnetic Bénard problem (8.1.5) when the Chadrasekhar number Q^2 is large. These are important physical problems and so an accurate non-linear theory is desirable. It has been possible to develop generalised energy theories (employing diverse Lyapunov functions) to derive non-linear energy stability results which yield Rayleigh number thresholds close to the instability thresholds delivered by linear stability theory, see e.g. the accounts in Straughan (1992). However, at the time of writing, the stability so obtained is conditional in that it only holds for a restricted class of initial data which can become vanishingly small. It is an important problem to extend the non-linear energy stability theory to be applicable for all initial data and so derive an unconditional stability result, but simultaneously maintaining the non-linear stability Rayleigh number threshold such that it is physically useful.

In general, genuinely unconditional non-linear energy stability results are few for problems where the non-linearities are not simply the convective ones, or where the linear operator $\mathcal{L} = L + \epsilon M$ is far from being symmetric. Such results have been obtained in penetrative convection, Payne & Straughan (1987), in anisotropic penetrative convection in a porous medium, Straughan & Walker (1996a), in thermal convection with black body radiation, Neitzel et al. (1994), in convection in a dielectric fluid fluid layer, Mulone et al. (1996), in convection in a vertical column of porous material, Straughan (1988), and in thawing subsea permafrost, see e.g. Straughan (1992), Chap. 7. However, these results are achieved by a variety of ad hoc techniques requiring some care in selection of a generalised energy, or Lyapunov functional.

8.1.1 Parallel Flows

Consider a viscous incompressible fluid contained in the infinite three - dimensional spatial layer I given by $y \in (-1, 1)$, $(x, z) \in \mathbf{R}^2$. One of the areas of hydrodynamic stability where the non-linear theory of energy stability has been least successful is in the study of parallel flows in the domain I, e.g. in the study of the stability of flows with a base solution like $\mathbf{v} = (U(y), 0, 0)$. Typical of such flows are Couette flow where $U(y) = y$ and Poiseuille flow for which $U(y) = 1 - y^2$. Couette flow is that which arises when the top plate $y = 1$ is sheared at a constant velocity relative to the bottom one, whereas Poiseuille flow is achieved by application of a constant pressure gradient in the x-direction, keeping the planes $y = \pm 1$ fixed.

The theory of the above flows is governed by the Navier–Stokes equations for the velocity and pressure fields, v_i, p. These partial differential equations are, in a suitably non-dimensionalised form,

$$\frac{\partial v_i}{\partial t} + Re\, v_j \frac{\partial v_i}{\partial x_j} = -\frac{\partial p}{\partial x_i} + \Delta v_i,$$

$$\frac{\partial v_i}{\partial x_i} = 0,$$

(8.1.6)

where $Re\, (= V\hat{L}/\nu)$ is the Reynolds number. (The quantities $V, \hat{L}$ and ν are a typical velocity, depth of the layer before non-dimensionalisation, and the kinematic viscosity of the fluid.) To study the stability of Couette or Poiseuille flow one may set $\mathbf{U} = (U(y), 0, 0)$ and then derive equations for the perturbation velocity and pressure fields (u_i, π) defined by

$$v_i = U_i + u_i, \qquad p = \bar{p} + \pi,$$

where $\bar{p}$ is the pressure corresponding to the base velocity U_i. The perturbation velocity and pressure u_i and π then satisfy the partial differential equations

$$\frac{\partial u_i}{\partial t} + Re\left(U \frac{\partial u_i}{\partial x} + \delta_{i1} U' v\right) + Re\, u_j \frac{\partial u_i}{\partial x_j} = -\frac{\partial \pi}{\partial x_i} + \Delta u_i,$$

$$\frac{\partial u_i}{\partial x_i} = 0,$$

(8.1.7)

where $U' = dU/dy$ and where $\mathbf{u} = (u, v, w)$.

The classical theory of linear instability for the system of partial differential equations (8.1.7) writes the functions u_i and π in the form

$$u_i = u_i(y) e^{i(ax + bz - act)},$$

$$\pi = \pi(y) e^{i(ax + bz - act)},$$

and discards the quadratic term $Re\,u_j\partial u_i/\partial x_j$ in $(8.1.7)_1$. This leads to the following system of ordinary differential equations

$$
\begin{aligned}
(ReU - c)iau + ReU'v &= -ia\pi + (D^2 - [a^2 + b^2])u, \\
(ReU - c)iav &= -D\pi + (D^2 - [a^2 + b^2])v, \\
(ReU - c)iaw &= -ib\pi + (D^2 - [a^2 + b^2])w, \\
iau + Dv + ibw &= 0,
\end{aligned}
\tag{8.1.8}
$$

where $D = d/dy$ and $\mathbf{u}(y) = (u(y), v(y), w(y))$. The traditional approach at this point has been to invoke Squire's theorem, arguing that the transformations

$$
Re \to \tilde{Re}\,\frac{\tilde{a}}{a}, \qquad c \to \tilde{c}\,\frac{\tilde{a}}{a}, \qquad \tilde{a} = \sqrt{a^2 + b^2},
$$

reduce (8.1.8) to a two-dimensional form

$$
\begin{aligned}
(ReU - c)iau + ReU'v &= -ia\pi + (D^2 - a^2)u, \\
(ReU - c)iav &= -D\pi + (D^2 - a^2)v, \\
iau + Dv &= 0.
\end{aligned}
\tag{8.1.9}
$$

The no-slip boundary condition at the plates is interpreted mathematically by requiring $u = v = 0$ for $y = \pm 1$. Next introduce a stream function ψ by

$$
u = \frac{\partial \psi}{\partial y}, \qquad v = -\frac{\partial \psi}{\partial x},
\tag{8.1.10}
$$

and then introduce the function $\phi(y)$ by

$$
\psi = \phi(y)\,e^{ia(x-ct)}.
\tag{8.1.11}
$$

In this manner, one shows from (8.1.9)–(8.1.11) that the linear instability problem reduces to studying the fourth order ordinary differential equation for $\phi(y)$,

$$
(D^2 - a^2)^2\phi = iaRe(U - c)(D^2 - a^2)\phi - iaReU''\phi, \qquad y \in (-1, 1). \tag{8.1.12}
$$

This is the celebrated Orr–Sommerfeld equation. The boundary conditions become

$$
\phi = D\phi = 0, \qquad y = \pm 1.
\tag{8.1.13}
$$

Equation (8.1.12) subject to the boundary conditions (8.1.13) constitutes an eigenvalue problem for the growth rate $c = c_r + ic_i$. If $c_i > 0$ the flow is linearly unstable. The solution to (8.1.12) subject to the boundary conditions (8.1.13) for the spectrum $\{c^{(k)}\}$ is a hard numerical problem, see e.g. the exposition in Dongarra *et al.* (1996).

Recently, Butler & Farrell (1992), Gustavsson (1991), and Reddy & Henningson (1993) have given convincing arguments to assert that the transient onset of instability is not governed solely by the leading eigenvalue of (8.1.12),

(8.1.13). In other words, the stability of flows such as Couette and Poiseuille are not completely controlled by the dominant eigenfunction $\phi^{(1)}$ for which $c^{(1)}$ has largest imaginary part. Indeed, they argue that one really ought to consider the full three-dimensional (linear) system (8.1.8).

In the interests of clarity we rederive the three-dimensional system in the manner of Butler & Farrell (1992). Equations (8.1.6) are

$$
\begin{aligned}
&\frac{\partial u}{\partial t} + Re\left(U\frac{\partial u}{\partial x} + U'v\right) = -\frac{\partial \pi}{\partial x} + \Delta u, \\
&\frac{\partial v}{\partial t} + ReU\frac{\partial v}{\partial x} = -\frac{\partial \pi}{\partial y} + \Delta v, \\
&\frac{\partial w}{\partial t} + ReU\frac{\partial w}{\partial x} = -\frac{\partial \pi}{\partial z} + \Delta w, \\
&\frac{\partial u}{\partial x} + \frac{\partial v}{\partial y} + \frac{\partial w}{\partial z} = 0.
\end{aligned}
\tag{8.1.14}
$$

Butler & Farrell (1992) argue that it is sufficient to consider the component of velocity normal to the plane of the flow, i.e. v, and the component of vorticity in the normal (y) direction. The normal component of vorticity is $\omega_2 = (\mathrm{curl}\,\mathbf{u})_2$ which may be written as ω, and then

$$
\omega = \frac{\partial u}{\partial z} - \frac{\partial w}{\partial x}.
\tag{8.1.15}
$$

Equipped with a knowledge of v and ω one may then calculate the variables u, w and π. The physical reason for v and ω being the main variables to influence the instability process is that a streamwise vortex is seen to be influential in experiments. From a mathematical viewpoint, v and ω arise naturally in the kinetic energy.

Butler & Farrell (1992) employ a different non-dimensionalisation which writes (8.1.14)$_{1-3}$ as

$$
\frac{\partial u_i}{\partial t} + U\frac{\partial u_i}{\partial x} + U'v\delta_{1i} = -\frac{\partial \pi}{\partial x_i} + \frac{1}{Re}\Delta u_i.
\tag{8.1.16}
$$

The evolution equation for ω is found directly from (8.1.16) by differentiating the equation for u with respect to z and subtracting from this the x derivative of the equation for w. The result is

$$
\frac{\partial \omega}{\partial t} + U\frac{\partial \omega}{\partial x} - \frac{1}{Re}\Delta\omega = -U'\frac{\partial v}{\partial z}.
\tag{8.1.17}
$$

A single equation for v is obtained from (8.1.16) by taking curlcurl of that equation to find

$$
\begin{aligned}
&-\Delta\frac{\partial u_i}{\partial t} + \frac{\partial^2}{\partial x_i\partial x_j}\left(U\frac{\partial u_j}{\partial x}\right) - \Delta\left(U\frac{\partial u_i}{\partial x}\right) \\
&+ \delta_{j1}\frac{\partial^2}{\partial x_i\partial x_j}(U'v) - \Delta(U'v)\delta_{i1} = -\frac{1}{Re}\Delta^2 u_i.
\end{aligned}
$$

Upon taking the second component of this equation one derives

$$\frac{\partial}{\partial t}\Delta v + U\frac{\partial}{\partial x}\Delta v - U''\frac{\partial v}{\partial x} - \frac{1}{Re}\Delta^2 v = 0. \tag{8.1.18}$$

Equations (8.1.17) and (8.1.18) are the coupled system of equations for v and w derived and used by Butler & Farrell (1992). The boundary conditions for v and w are that

$$v = \frac{\partial v}{\partial y} = w = 0, \qquad \text{at } y = \pm 1.$$

Butler & Farrell (1992) write v and w in the form (which is equivalent to that used before (8.1.8)),

$$\begin{aligned} v &= v(y)\,e^{i(ax+bz)+\sigma t}, \\ w &= w(y)\,e^{i(ax+bz)+\sigma t}. \end{aligned} \tag{8.1.19}$$

Representation (8.1.19) employed in the partial differential equations (8.1.17) and (8.1.18) leads to the following system of ordinary differential equations for v and w,

$$\begin{aligned} (D^2 - k^2)^2 v - iaReU(D^2 - k^2)v + iaReU''v &= Re\sigma(D^2 - k^2)v, \\ (D^2 - k^2)w - iaReUw - ibReU'v &= Re\sigma w, \end{aligned} \tag{8.1.20}$$

where $k^2 = a^2 + b^2$. System (8.1.20) is to be solved on the interval $y \in (-1, 1)$ subject to the boundary conditions

$$v = Dv = 0, \quad w = 0, \qquad y = \pm 1. \tag{8.1.21}$$

We shall refer to (8.1.20), (8.1.21) as the Butler–Farrell eigenvalue problem. The eigenvalues are $\sigma^{(i)}$ with eigenfunctions $\{v^{(i)}, w^{(i)}\}$.

We are now in a position to return to a discussion of stability of Couette and Poiseuille flows. For Couette flow the linearised theory of instability based on equation (8.1.12) and the boundary conditions (8.1.13) predicts the flow is always stable, i.e. $c_i < 0$ for all eigenvalues. For Poiseuille flow linear theory based on (8.1.12) and (8.1.13) yields instability for $Re > 5772.22$, see e.g. Orszag (1971). The critical wavenumber for instability is $a_c = 1.02056$, Orszag (1971). When one bases a *non-linear energy stability* theory on the kinetic energy then unconditional (i.e. for all initial data) non-linear stability is found for Couette flow when $Re < 20.7$, whereas non-linear stability follows for Poiseuille flow when $Re < 49.6$, cf. Joseph (1976). Experimental work, on the other hand, has visualised instabilities for Reynolds numbers of the order of 1000. Thus, the linear instability theory is of little use in predicting accurately the onset of instability since the values of $Re = \infty$ and $Re = 5772.22$ are too large. Energy stability theory, on the other hand, is far too conservative in the stability boundary it yields. This is a case where the non-linear energy method has not to date proved too useful.

In an interesting paper, Rionero & Mulone (1991) have employed a generalised energy of form

$$V(t) = \frac{1}{2}(\|\mathbf{j} \cdot \nabla \times \mathbf{u}\|^2 + \beta\|\Delta v\|^2) + \frac{1}{2}\mu(\|\nabla\mathbf{u}\|^2 + \|\nabla \cdot (\nabla \times \mathbf{u})\|^2)$$

where $\mathbf{j} = (0, 1, 0)$, and β and μ are coupling parameters selected optimally. They show that Couette or Poiseuille flow is non-linearly stable for all Reynolds numbers, in V measure, for *stress free boundary conditions*. Their result is a conditional one with the size of the initial data being restricted. Due to the initial data restriction this result may also be interpreted as a rigorous linearisation principle. However, I am unaware of any non-linear energy stability results for the *fixed surface problem* which yield Reynolds numbers close to those observed in experiments.

The purpose of this chapter is to focus on work which has observed rapid growth of the kinetic energy associated with the linearised version of (8.1.6). These interesting developments and ramifications are given in Berlin *et al.* (1994), Butler & Farrell (1992, 1993), Farrell (1988a, 1988b, 1989, 1990), Farrell & Ioannou (1993a, 1993b, 1993c, 1994a), Gustavsson (1991), Henningson (1995), Henningson *et al.* (1993), Henningson & Reddy (1994), Hooper & Grimshaw (1995), Kreiss *et al.* (1994), Reddy & Henningson (1993), Reddy *et al.* (1993), Schmid & Henningson (1994), and the references therein.

To explain why instability in Couette or Poiseuille flow is seen in practice at much lower Reynolds numbers than those predicted by classical linear instability theory, investigations by Mack (1976), Gustavsson (1981, 1986), Gustavsson & Hultgren (1980), and Shanthini (1989), studied the spectrum of the Orr–Sommerfeld operator to see if resonances between eigenvalues could be responsible. A resonance occurs where an eigenvalue is exactly repeated. In that case one of the eigenfunctions of the repeated eigenvalue contains a linear t growth term. (The situation is analogous to the well known case of a second order ordinary differential equation with constant coefficients. When there are repeated roots of the auxilliary equation one of the eigenfunctions grows linearly in t.) This could conceivably lead to strong transient algebraic growth of the perturbation at short times which is eventually damped out exponentially according to linear theory. While several resonances were found in numerical studies, no substantial growth would appear to have been predicted: the comments of Butler & Farrell (1992), p. 1645, on this matter are very pertinent. (A resonance in the numerical sense was interpreted as two eigenvalues being a pre-requested distance apart in the complex plane.) The idea that resonances could be responsible for transient solution growth means that it is not sufficient to investigate only the eigenvalue (and eigenfunction) of the Orr–Sommerfeld problem which has greatest imaginary part. This has lead to the development of numerical methods which can accurately yield *all* the eigenvalues and eigenfunctions of (8.1.12) and (8.1.13), or at least we find sufficient eigenvalues at the "top end" of the spectrum. By the "top end" of the spectrum for system (8.1.12) and (8.1.13) we mean those eigenvalues which have largest imaginary parts. We typically calculate all eigenvalues for which $c_i > -1$.

In Figs. 8.1 and 8.2 below we show the top end of the spectrum for (8.1.12), (8.1.13) in the case of Poiseuille flow with $Re = 5772.22$, $a = 1.02056$ (Fig. 8.1), and for Couette flow with $Re = 900$, $a = 1.2$ (Fig. 8.2). The even eigenfunctions satisfy $\phi(y) = \phi(-y)$ whereas the odd ones are such that $\phi(-y) = -\phi(y)$.

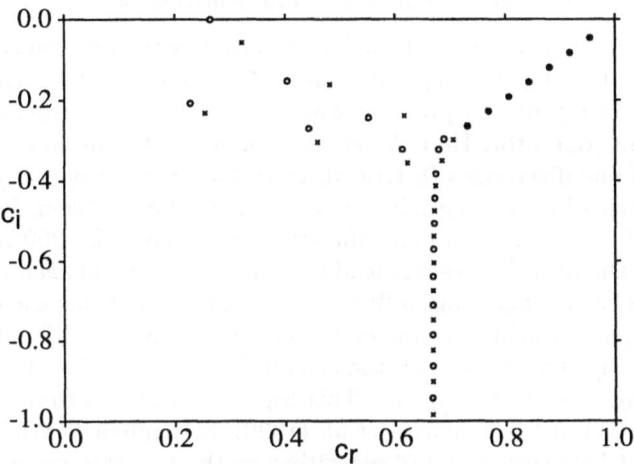

Fig. 8.1 The spectrum for plane Poiseuille flow, $U = 1 - y^2$, with $Re = 5772.22$, $a = 1.02056$. The open circles represent even eigenmodes, crosses represent odd ones.

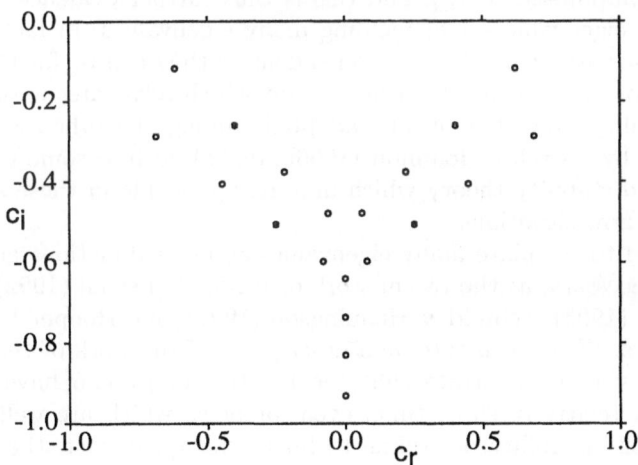

Fig. 8.2 The spectrum for Couette flow, $U = y$, with $Re = 900$, $a = 1.2$.

The eigenvalues in Fig. 8.1 extend downward with $c_r \to 2/3$ and form a countably infinite set. Those in Fig. 8.2 do likewise with $c_r = 0$. The eigenvalues in Figs. 8.1 and 8.2 are calculated using the technique advocated by Dongarra *et al.* (1996). These writers write the differential equation (8.1.12) as a system of two second-order equations,

$$(D^2 - a^2)\phi - \chi = 0,$$
$$(D^2 - a^2)\chi - iaRe(U - c)\chi + iaReU''\phi = 0. \tag{8.1.22}$$

Dongarra *et al.* (1996) solve (8.1.22) by a Chebyshev tau method coupled with the QZ algorithm. The beauty of dealing with system (8.1.22) is that the generalised matrix eigenvalue problem which arises contains terms like $O(M^3)$ due to the D^2 operator. Here M is the number of Chebyshev polynomials employed. If one discretizes (8.1.12) directly then the D^4 operator gives rise to $O(M^7)$ terms in the generalised matrix eigenvalue problem. For accurate resolution of eigenvalues and eigenfunctions one may need 200 polynomials or more and the $O(M^7)$ case can lead to a loss of numerical accuracy. In fact, in both the Couette and Poiseuille flow eigenvalue problems, the eigenvalues and eigenfunctions near the branch of the "Y-shape" are difficult to calculate numerically. This is because the eigenfunctions of nearby eigenvalues are close to being linearly dependent. This topic is addressed from a numerical viewpoint in detail by Dongarra *et al.* (1996). Straughan & Walker (1996b) develop the Chebyshev tau–QZ algorithm method in the context of stability and instability for porous media convection problems. Farrell (1988a) has shown that for Poiseuille flow the symmetric eigenfunctions corresponding to those eigenvalues nearest the branch of the "Y" contribute most to the growth in time of the kinetic energy, and this is thus further evidence for needing an accurate eigenvalue solver yielding many eigenvalues. In fact, when the eigenfunctions are nearly linearly dependent as they can be for Couette and Poiseuille flow there is a school of thought which advocates another definition of stability from that usually adopted, see e.g. Trefethen *et al.* (1992). The papers by Farrell & Ioannou (1996b, 1996c) go into some detail about a generalised stability theory which may be applicable in these and related geophysical flow situations.

The need to calculate many eigenvalues and eigenfunctions accurately is now very necessary, as the recent work of Butler & Farrell (1992), Reddy & Henningson (1993), Schmid & Henningson (1994), and Hooper & Grimshaw (1995) shows. We concentrate on a description of the work of Butler & Farrell (1992) who demonstrate that the kinetic energy can have very large growth in a relatively short time, even for flows which are well below the threshold for instability according to linear theory. In fact, the variational method employed by Butler & Farrell (1992) is based on earlier work of Farrell (1988a, 1988b, 1989) on growth in various two-dimensional geophysical fluid flows.

Butler & Farrell (1992) consider the kinetic energy of a perturbation to Couette or Poiseuille flow, i.e. they study the function

$$E(t) = \frac{1}{2V} \int_{-1}^{1} \int_{0}^{A} \int_{0}^{B} (u^2 + v^2 + w^2)\, dz\, dx\, dy, \qquad (8.1.23)$$

where A and B are the x and z wavelengths and V is the volume of the energy cell. Due to the fact that the system is linearised they can work with v and w as defined in (8.1.19) which they also write as

$$v(x, y, z, t) = \hat{v}(y, t)\, e^{i(ax+bz)},$$
$$w(x, y, z, t) = \hat{w}(y, t)\, e^{i(ax+bz)}, \qquad (8.1.24)$$

where $\hat{v}, \hat{w}$ can be complex, although only the real parts are used in (8.1.24). Butler & Farrell (1992) show that the kinetic energy (8.1.23) may be written in terms of v and w as

$$E(t) = \frac{1}{8} \int_{-1}^{1} \left[\hat{v}^* \hat{v} + \frac{1}{k^2} \left(\frac{\partial \hat{v}^*}{\partial y} \frac{\partial \hat{v}}{\partial y} + \hat{w}^* \hat{w} \right) \right] dy, \qquad (8.1.25)$$

where a $*$ denotes complex conjugate and $k^2 = a^2 + b^2$. To compute numerical calculations Butler & Farrell (1992) discretize v and w by writing

$$v = \sum_{j=1}^{2N} \gamma_j \left[\tilde{v}_j \exp(\sigma_j t) \right] \exp \left[i(ax + bz) \right],$$
$$w = \sum_{j=1}^{2N} \gamma_j \left[\tilde{w}_j \exp(\sigma_j t) \right] \exp \left[i(ax + bz) \right], \qquad (8.1.26)$$

where γ_j is the spectral projection on the jth mode of the Butler–Farrell eigenvalue problem (8.1.20), (8.1.21). The technique of Butler & Farrell (1992) to find the eigenfunction and eigenvalue of the jth mode is to discretize using finite differences and then employ the QR algorithm on the generalised matrix eigenvalue problem. The techniques of Reddy & Henningson (1993) and Hooper & Grimshaw (1995) also find energy growth although the numerical method underpinning the work of Reddy & Henningson (1993) is a Chebyshev collocation one, whereas Hooper & Grimshaw (1995) employ a Chebyshev tau technique. Hence, Butler & Farrell (1992) adopt the notation

$$v \equiv V_{mj} \gamma_j e^{i(ax+bz)}, \qquad w \equiv \Omega_{mj} \gamma_j e^{i(ax+bz)},$$

where

$$V_{mj} = \tilde{v}_{mj} e^{\sigma_j t}, \qquad \Omega_{mj} = \tilde{w}_{mj} e^{\sigma_j t},$$

with m denoting a value between 1 and N and referring to the finite difference point $y_{m+1} = m\Delta y$ in the interval $y \in (-1, 1)$. This allows Butler & Farrell (1992) to approximate the energy (8.1.25) by a finite dimensional form

$$E(t) = \frac{\Delta y}{8} \left[\gamma_p^* V_{mp}^* V_{mj} \gamma_j + \frac{1}{k^2} \left\{ \gamma_p^* \frac{\partial V_{mp}^*}{\partial y} \frac{\partial V_{mj}}{\partial y} \gamma_j + \gamma_p^* \Omega_{mp}^* \Omega_{mj} \gamma_j \right\} \right], \quad (8.1.27)$$

where summation over the various subscripts is understood. The form of $E(t)$ is conveniently rewritten, Butler & Farrell (1992), as

$$E(t) = \gamma_j^* E_{ji}(t) \gamma_i. \quad (8.1.28)$$

The form for the Hermitian matrix E_{ji} may be found from (8.1.27) and the time dependence of E_{ij} has been explicitly pointed out.

The idea of Butler & Farrell (1992) is to fix the wavenumbers a and b, fix the Reynolds number Re, and then find the linear perturbation which maximises $E(t)$ at time t subject to the constraint that the initial energy has the numerical value of 1. This yields a maximisation problem for the function F given by

$$F = \gamma_j^* E_{ji}(t) \gamma_i + \lambda \left(\gamma_j^* E_{ji}(0) \gamma_i - 1 \right),$$

for some Lagrange multiplier λ. The solution to this maximisation problem is found from the Euler–Lagrange equations

$$E_{ij}(t) \gamma_j + \lambda E_{ij}(0) \gamma_j = 0. \quad (8.1.29)$$

The eigenvalues λ represent the ratio $E(t)/E(0)$ for an eigenvector γ_i. (The technical details of a practical way to do this calculation using the Chebyshev tau method are given in the very readable account of Hooper & Grimshaw (1995).) The calculation of λ involves finding $E_{ij}(t)$ and this in turn involves calculation of the eigenvalues and eigenfunctions of the Butler–Farrell eigenvalue problem (8.1.20), (8.1.21).

Butler & Farrell (1992) have completed extensive calculations of solutions to (8.1.29) and have computed the growth rate for many situations in Couette flow, in Poiseuille flow, and even in Blasius flow.

For Couette flow, Butler & Farrell (1992) find with $Re = 1000$ that the global optimal for time $\tau = 117$ units is achieved with $a = 0.035$ and $b = 1.60$. This yields an energy ratio of $E(\tau)/E(0) = 1185$. This is certainly an impressive growth of the kinetic energy. Since only linear theory is considered, the linear energy eventually decays, as shown schematically in Fig. 8.3.

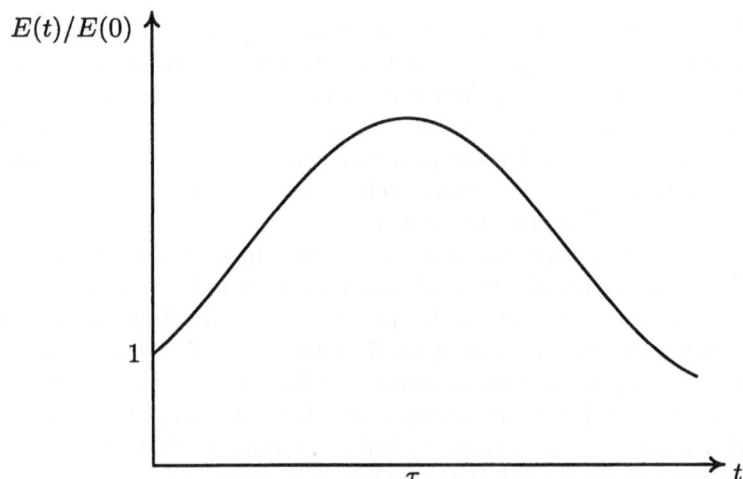

Fig. 8.3 A schematic picture of the energy growth in time for Couette or Poiseuille flow.

We stress that Butler & Farrell (1992) show that a two-dimensional perturbation yields an energy growth of $O(13)$ and thus the perturbation causing largest energy disturbance is truly three-dimensional. Similar results pertaining to energy growth are found by Reddy & Henningson (1993), Hooper & Grimshaw (1995), for Couette and Poiseuille flow, and by Schmid & Henningson (1994) for the problem of Poiseuille flow in a circular pipe. An expanded exposition of the last named problem is given in a subsection after this.

For the problem of Poiseuille flow, Butler & Farrell (1992) obtain similar behaviour to that of Fig. 8.3 for growth of the kinetic energy. For $Re = 5000$, a value which is stable according to classical linear instability theory, they determine optimal energy growth at time $\tau = 379$ units. The respective wavenumbers are $a = 0$ and $b = 2.044$. The energy ratio is $E(\tau)/E(0) = 4897$.

The work of Butler & Farrell (1992), Reddy & Henningson (1993) and Schmid & Henningson (1994) is an exciting development of rapid solution growth in three-dimensional flows which have traditionally proved hard to analyse. When the effects of non-linearity are fully understood and added to this it could well describe quantitatively the formation of patches of turbulent fluid flow. In the context of fully non-linear flow, Butler & Farrell (1994) have analysed the non-linear development of two-dimensional perturbations which employ the optimal configuration of the linear problem, namely that which gains most energy. To do this they employ a vorticity streamfunction numerical simulation using a Fourier spectral discretization in x with finite differences in y. These results are very interesting and display finite amplitude solutions which persist for a long time. In connection with this it is worth drawing attention to the fact that Farrell & Ioannou

(1993a, 1993b, 1993c, 1994a) have presented interesting analyses of the effects of stochastic forcing of the equations of Couette and Poiseuille flow, while Farrell & Ioannou (1993d, 1994b) derive stimulating stochastic and statistical results for baroclinic waves. Farrell & Ioannou (1995) produces highly useful stochastic dynamics and conclusions for the midlatitude atmospheric jet; one of the ultimate goals of a study such as the last one is to produce an understanding of the global climate system.

It is also pertinent at this juncture to mention further work on energy growth and stochastic dynamics by Professor Farrell and his co-workers. Farrell & Ioannou (1996a) investigate the problem of controlling the energy growth and stochastic dynamics of shear flow by controlling the boundary conditions. This is very interesting because it yields information on how one may control the transistion to turbulence and how one may, therefore, be able to possibly suppress turbulence. Farrell & Ioannou (1997) analyses the shear flow problem further examining especially the strong shear effects which occur in the boundary layers. They are able to develop a stochastic analysis to study the boundary layer frequency wavenumber spectra. DelSole & Farrell (1995, 1996) study in depth the dynamical properties of a stochastically excited two layer baroclinic flow which is damped by thermal effects. The dynamics are described by streamfunctions measuring the average and difference of the stream function in the two layers. Moore & Farrell (1994) studies error growth in flows related to ocean modelling. This paper contains many other relevant references. They explain how to study error growth in complex ocean flows and relate this to predictability studies in meteorology.

An interesting technique for finding a bound for the solution of turbulent shear flow is discussed in the book of Doering & Gibbon (1995). These writers include an exposition of a method which decomposes the solution to the shear flow problem into a background flow and a perturbation. The background flow incorporates the non-homogeneous boundary conditions. A sharp resolution of the Euler–Lagrange equations associated with the background decomposition technique has recently been provided in the highly interesting paper by Kerswell (1997).

8.1.2 Energy Growth in Circular Pipe Flow

The paper by Schmid & Henningson (1994) develops an analysis in a somewhat similar vein to that of Butler & Farrell (1992), but for the problem of flow in a circular pipe (Hagen–Poiseuille flow). The technical details of the method employed by Schmid & Henningson (1994) are different from those of Butler & Farrell (1992), and this paper like that of Henningson & Reddy (1993) is worth reading in its own right.

In cylindrical polar coordinates (r, θ, z), the equations for a perturbation to the axial mean flow $U = U(r)$, whose instability is under investigation, have form

$$\frac{\partial u}{\partial t} + U\frac{\partial u}{\partial z} + U'v = -\frac{\partial p}{\partial z} + Re^{-1}\left[\frac{1}{r}\frac{\partial}{\partial r}\left(r\frac{\partial u}{\partial r}\right) + \frac{1}{r^2}\frac{\partial^2 u}{\partial\theta^2} + \frac{\partial^2 u}{\partial z^2}\right],$$

$$\frac{\partial v}{\partial t} + U\frac{\partial v}{\partial z} = -\frac{\partial p}{\partial r} + Re^{-1}\left[\frac{1}{r}\frac{\partial}{\partial r}\left(r\frac{\partial v}{\partial r}\right) + \frac{1}{r^2}\frac{\partial^2 v}{\partial\theta^2} + \frac{\partial^2 v}{\partial z^2}\right.$$
$$\left. -\frac{v}{r^2} - \frac{2}{r^2}\frac{\partial w}{\partial\theta}\right],$$

$$\frac{\partial w}{\partial t} + U\frac{\partial w}{\partial z} = -\frac{1}{r}\frac{\partial p}{\partial\theta} + Re^{-1}\left[\frac{1}{r}\frac{\partial}{\partial r}\left(r\frac{\partial w}{\partial r}\right) + \frac{1}{r^2}\frac{\partial^2 w}{\partial\theta^2}\right.$$
$$\left. + \frac{\partial^2 w}{\partial z^2} + \frac{2}{r^2}\frac{\partial v}{\partial\theta} - \frac{w}{r^2}\right],$$

$$\frac{\partial u}{\partial z} + \frac{1}{r}\frac{\partial(rv)}{\partial r} + \frac{1}{r}\frac{\partial w}{\partial\theta} = 0.$$

(8.1.30)

Here u, v, w are the perturbation velocities in the z, r and θ directions and Re is the Reynolds number. Schmid & Henningson (1994) write u, v, w and the pressure perturbation p in the form

$$\begin{pmatrix} u \\ v \\ w \\ p \end{pmatrix} = \begin{pmatrix} \hat{u} \\ \hat{v} \\ \hat{w} \\ \hat{p} \end{pmatrix} e^{(i\alpha z + n\theta)}$$

for some real number α, with n an integer, and then they work with the radial velocity $\hat{v}$ and a radial vorticity $\hat{\eta}$, cf. the analysis of Butler & Farrell (1992). In terms of the variables Φ and Ω given by

$$\Phi = -ir\hat{v}, \qquad \Omega = \frac{\alpha r\hat{w} - n\hat{u}}{nRek^2r^2} = \frac{\hat{\eta}}{inRek^2r},$$

where

$$k^2 = \alpha^2 + \frac{n^2}{r^2},$$

the perturbation equations become, Schmid & Henningson (1994),

$$Re\left(\frac{\partial}{\partial t} + i\alpha U\right)T\Phi - \frac{i\alpha Re}{r}\left(\frac{U'}{k^2r}\right)'\Phi = T(k^2r^2T)\Phi + 2\alpha Re\, n^2 T\Omega,$$

$$k^2r^2Re\left(\frac{\partial}{\partial t} + i\alpha U\right)\Omega + \frac{iU'}{r}\Phi = S\Omega + \frac{2\alpha}{R}T\Phi,$$

(8.1.31)

where the operators T and S have form

$$T = \frac{1}{r^2} - \frac{1}{r}\frac{\partial}{\partial r}\left(\frac{1}{k^2r}\frac{\partial}{\partial r}\right),$$

and

$$S = k^4 r^2 - \frac{1}{r}\frac{\partial}{\partial r}\left(k^2 r^3 \frac{\partial}{\partial r}\right).$$

As Schmid & Henningson (1994) show, $\hat{u}$ and $\hat{w}$ may be recovered from a knowledge of Φ and Ω.

The boundary conditions are

$$\Phi = \Phi' = \Omega = 0, \qquad \text{at } r = 1,$$

while those for Φ and Ω at the axis $r = 0$ depend on what value of n is being considered.

Schmid & Henningson (1994) develop an energy growth analysis in the measure

$$E = \pi \int_0^1 (|\hat{u}|^2 + |\hat{v}|^2 + |\hat{w}|^2) r\, dr$$

$$= \pi \int_0^1 \left(\frac{1}{k^2 r^2}|\hat{\Phi}'|^2 + \frac{1}{r^2}|\hat{\Phi}|^2 + k^2 r^2 n^2 Re^2 |\hat{\Omega}|^2\right) r\, dr.$$

The analysis of energy growth in Schmid & Henningson (1994) is developed by solving the eigenvalue problem which arises by a Chebyshev collocation technique. The spectrum for $n = 0$, $Re = 5000$, $a = 1.1$ is given in Fig. 8.4. (The spectrum as shown in this figure has been computed using the prescription given in Dongarra et al. (1996).)

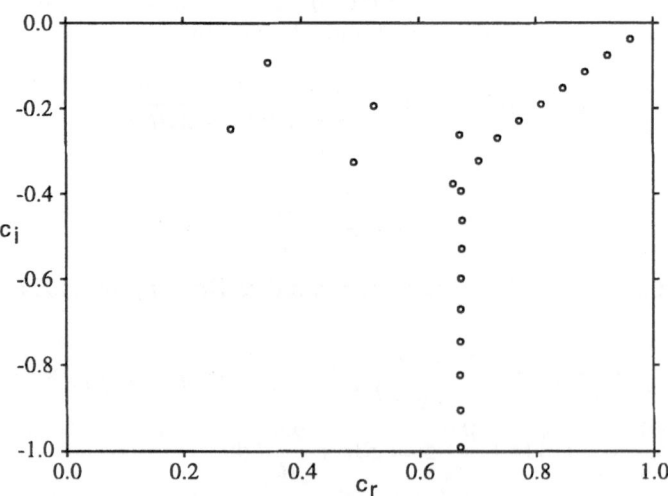

Fig. 8.4 The spectrum for Hagen–Poiseuille flow with $n = 0$, $Re = 5000$, and $a = 1.1$.

Schmid & Henningson (1994) include a picture of the spectrum for $n = 1$, and they demonstrate where the cases $n \neq 0$ are important. Again they find large transient growth is possible as in the case of plane Poiseuille flow.

The paper of Schmid and Henningson (1994), pp. 212, 213, contains an interesting account of how the numerical range of the appropriate operator is relevant, why pseudospectra could be important, and they relate this to the Hille–Yosida theorem.

8.1.3 Linear Instability of Elliptic Pipe Flow

The problem of instability of flow in an elliptic pipe has attracted the attention of many prominent writers. A beautiful analysis of this problem is contained in the paper by Kerswell & Davey (1996) where several other references may be found. This problem is of interest, among other things, because the solution to the plane Poiseuille flow instability problem becomes unstable according to linear theory for $Re \approx 5772.22$ whereas for axisymmetric disturbances Poiseuille flow in a pipe circular pipe is believed stable according to linear theory. By examining the limits of pipe aspect ratio $A \to \infty$ and $A \to 1$, the elliptic pipe flow problem encompasses both cases.

For a steady flow in an elliptical pipe with boundary satisfying the equation

$$\frac{x^2}{1 + \beta} + \frac{y^2}{1 - \beta} = 1,$$

with aspect ratio

$$A = \left(\frac{1 + \beta}{1 - \beta} \right)^{1/2},$$

the basic solution is

$$\mathbf{U} = \left(1 - \frac{x^2}{1 + \beta} - \frac{y^2}{1 - \beta} \right) \hat{\mathbf{z}}, \qquad (8.1.32)$$

where $\hat{\mathbf{z}}$ is the unit vector in the z-direction, see Kerswell & Davey (1996).

Kerswell & Davey (1996) develop a linear instability analysis for the solution (8.1.32) employing fully three-dimensional perturbations by working with the spatial variables (s, ϕ, z) given by

$$x = s\sqrt{1 + \beta} \cos \phi, \qquad y = s\sqrt{1 - \beta} \sin \phi.$$

The analysis of Kerswell & Davey (1996) is technical and we do not include details here. However, we stress that these writers introduce a very efficient method of solving the hydrodynamic stability eigenvalue problem which is based on a Chebyshev collocation technique. They work *directly* with the variables u, v, w and p and report that this method does not produce spurious eigenvalues. Indeed, I believe that this method is of much interest in its own right.

The paper of Kerswell & Davey (1996) is well worth reading and the numerical results indicate that the critical Reynolds number appears to fall from infinity at a value of A just above 10 and then approach the asymptotic result

$$Re_{\text{crit}} = 5772 + \frac{86300}{A},$$

as A increases.

I am unaware of an analysis of energy growth for the elliptic pipe problem, in the vein of those of Butler & Farrell (1992), Reddy & Henningson (1993) or Schmid & Henningson (1994).

8.2 Transient Growth in Compressible Flows

In a recent piece of work Hanifi et al. (1995) have developed an analysis of energy growth which is in some sense analogous to that of the last section, but for a layer of compressible fluid. Due to the fact that the fluid is compressible the analysis is applicable to boundary layer flows in a gas and therefore has to include non-isothermal effects.

The equations for flow in a compressible fluid, linearised about a steady mean flow $U(y)$ in the x-direction, may be written, Hanifi et al. (1995),

$$\frac{\partial u_i}{\partial t} + U_j \frac{\partial u_i}{\partial x_j} = -U_i'v - \frac{1}{\bar{\rho}\gamma M^2}\frac{\partial p}{\partial x_i} + \frac{1}{\bar{\rho}Re}\frac{\partial \sigma_{ij}}{\partial x_j}, \qquad (8.2.1)$$

$$\frac{\partial \rho}{\partial t} + U_j \frac{\partial \rho}{\partial x_j} = -\bar{\rho}'v - \bar{\rho}\frac{\partial u_i}{\partial x_i}, \qquad (8.2.2)$$

$$\frac{\partial \theta}{\partial t} + U_i \frac{\partial \theta}{\partial x_i} = -T'v - (\gamma - 1)T\frac{\partial u_i}{\partial x_i} + \frac{\gamma}{\bar{\rho}PrRe}\Theta$$

$$+ \gamma(\gamma - 1)\frac{M^2}{\bar{\rho}Re}\Phi, \qquad (8.2.3)$$

where the pressure p has form

$$p = T\rho + \bar{\rho}\theta, \qquad (8.2.4)$$

with $\bar{\rho}, T$ being the density and temperature in the steady state, $\mathbf{u} = (u, v, w)$ is the disturbance velocity, and ρ and θ are the disturbance density and temperature, respectively. The coefficients M, Re and Pr are the Mach number, Reynolds number, and the Prandtl number. The quantities σ_{ij}, Θ and Φ are defined by

$$\sigma_{ij} = \lambda\delta_{ij}\frac{\partial u_m}{\partial x_m} + \mu\left(\frac{\partial u_i}{\partial x_j} + \frac{\partial u_j}{\partial x_i}\right) + \frac{d\mu}{dT}(U_i'\delta_{j2} + U_j'\delta_{i2})\theta,$$

$$\Theta = 2\kappa'\theta' + \kappa''\theta + \kappa\Delta\theta,$$

$$\Phi = 2\mu\left(U_i'u_i' + U_i'\frac{\partial v}{\partial x_i}\right) + \frac{d\mu}{dT}U_i'U_i'\theta,$$

where a prime denotes $\partial/\partial y$ and μ, λ, κ and γ are the first and second coefficients of viscosity, the thermal conductivity, and the specific heat ratio.

Hanifi *et al.* (1995) remark that in the compressible flow case it is not clear what functional one should use to represent the "energy". They select the form

$$E(t) = \frac{1}{2}\int_V (Au_i^*u_i + B\rho^*\rho + C\theta^*\theta)dx, \qquad (8.2.5)$$

where V is a suitable volume, the star denotes complex conjugate, and A, B and C are positive coefficients to be selected. There is much leeway in selecting these coefficients and Hanifi *et al.* base their selection such that the compression work is eliminated. The energy growth analysis follows the lines of the incompressible pipe flow case of Schmid & Henningson (1994), although here the technical details are very different. Indeed, due to the compressibility, the calculation is very involved. The calculation of eigenvalues and eigenvectors necessitates a technique capable of handling the compressible flow equations. To this end, Hanifi *et al.* (1995) employ a Chebyshev collocation method, cf. Malik (1990). The spectrum of the compressible flow problem is investigated in some detail by Hanifi *et al.* (1995) and this exhibits some highly interesting behaviour. We do not describe in detail the findings of Hanifi *et al.* (1995), although we do point out that they observe very rapid "energy" growth in time according to linear theory, for compressible boundary layer flow. This energy growth can be observed in flow regions which are stable according to linear theory, cf. Fig. 4 of Hanifi *et al.* (1995). Several other references to instability in compressible boundary layers are quoted by Hanifi *et al.* (1995) who also integrate their work into the context of that by other writers.

8.3 Shear Flow in Granular Materials

A very practical example of where shear flows can have a pronounced effect on human life are the occurrences of landslides or avalanches. Hutter *et al.* (1987) discuss the applicability of using shear flows in granular media to model such issues. Since the partial differential equations governing the behaviour of granular media are complex and mathematically resemble those of compressible flow rather than incompressible flow, any stability analysis will inevitably be highly involved. Nevertheless, promising analyses of the linearised instability problem for shear flow in a granular material have been presented by Savage (1992) and by Babic (1993). These writers assume the granular medium occupies an infinite region in space. For example, Babic (1993) assumes the material occupies the whole of $\mathbf{R}^3$.

Schmid & Kytömaa (1994) again work in the background of an infinite spatial region, but their goal is to investigate the growth of the linearised solution in a manner somewhat analogous to that described in the last two sections. These writers consider a granular medium occupying the whole of the spatial region $\mathbf{R}^2$ and adopt a base flow which is linear in the y variable. This flow is in the x-direction, i.e. the base velocity $\mathbf{U}$ is

$$\mathbf{U}(y) = (\Gamma y, 0, 0).$$

There is much discussion in the literature over the form of the equations for description of a granular medium. It is not our concern to enter these discussions, although the introductions to the papers of Schmid & Kytömaa (1994) and Wang *et al.* (1996) do provide very readable reviews. The partial differential equations governing the behaviour of flow in a granular body given by Schmid & Kytömaa (1994) or Wang (1996) *et al.* have form

$$\frac{\partial \rho}{\partial t} + \frac{\partial}{\partial x_i}(\rho u_i) = 0, \tag{8.3.1}$$

$$\rho\left(\frac{\partial u_i}{\partial t} + u_j \frac{\partial u_i}{\partial x_j}\right) = -\frac{\partial}{\partial x_j}\sigma_{ij}, \tag{8.3.2}$$

$$\frac{3}{2}\rho\left(\frac{\partial T}{\partial t} + u_i \frac{\partial T}{\partial x_i}\right) = -\frac{\partial q_i}{\partial x_i} - \sigma_{ij}\frac{\partial u_i}{\partial x_j} - J. \tag{8.3.3}$$

In these equations ρ is the bulk density of the medium, u_i is the mean velocity, σ_{ij} is the stress tensor, T is the granular temperature, q_i is a thermal flux vector and J denotes a rate of dissipation of energy which arises because of collisions between the grains. The granular temperature is a quantity

$$\frac{1}{3}\langle u'^2 \rangle$$

where u' is the magnitude of the fluctuation about the local mean velocity and $\langle \cdot \rangle$ is an average. Constitutive equations for σ_{ij}, q_i and J are given as, Wang *et al.* (1996),

$$\sigma_{ij} = \left[\rho T(1 + 4\eta\nu g_0) - \eta\mu_b \frac{\partial u_k}{\partial x_k}\right]\delta_{ij}$$

$$- \left(\frac{2+\alpha}{3}\right)\left[\frac{2\mu}{\eta(2-\eta)g_0}\left(1 + \frac{8}{5}\nu\eta g_0\right)\left[1 + \frac{8}{5}\eta(3\eta-2)\nu g_0\right] + \frac{6}{5}\eta\mu_b\right]S_{ij},$$

$$q_i = -\frac{\lambda}{g_0}\left\{\left(1 + \frac{12}{5}\nu\eta g_0\right)\left[1 + \frac{12}{5}\eta^2(4\eta-3)\nu g_0\right]\right.$$

$$\left. + \frac{64}{25\pi}(41 - 33\eta)(\eta\nu g_0)^2\right\}\frac{\partial T}{\partial x_i}$$

$$- \frac{\lambda}{g_0}\frac{12}{5}\eta(\eta-1)(2\eta-1)\left(1 + \frac{12}{5}\nu\eta g_0\right)\frac{d}{d\nu}(\nu^2 g_0)\frac{T}{\nu}\frac{\partial\nu}{\partial x_i},$$

$$J = \frac{48}{\sqrt{\pi}}\eta(1-\eta)\frac{\rho_p\nu^2}{d}g_0 T^{3/2},$$

where the tensor S_{ij} has form

$$S_{ij} = \frac{1}{2}\left(\frac{\partial u_i}{\partial x_j} + \frac{\partial u_j}{\partial x_i}\right) - \frac{1}{3}\frac{\partial u_k}{\partial x_k}\delta_{ij}.$$

The coefficients μ, μ_b and λ depend on the mass, M, and diameter, d, of the particles in the granular medium and g_0 is a constitutive function of ν chosen by Wang *et al.* (1996) to have form

$$g_0(\nu) = \frac{1}{1 - (\nu/\nu_m)^{1/3}}.$$

In this expression ν_m is the volume of solids in a closest packing configuration.

Clearly the system of partial differential equations is a formidable one and exhibits compressibility. Nevertheless, Schmid & Kytömaa (1994) develop a two-dimensional analysis by studying a two-space-dimensional perturbation (ν, u, v, T) about the steady shear $\mathbf{U}(y) = (\Gamma y, 0, 0)$. They work with perturbations in the form

$$\begin{pmatrix} \nu \\ u \\ v \\ T \end{pmatrix} = \begin{pmatrix} \hat{\nu} \\ \hat{u} \\ \hat{v} \\ \hat{T} \end{pmatrix} \exp\left[i(k_x x + k_y y)\right],$$

for a wave vector of form

$$\mathbf{k}(t) = \left(k_x(0), k_y(0) - tk_x(0)\right).$$

This allows them to obtain ordinary differential equations for the evolutionary behaviour of the quantities $\hat{\nu}, \hat{u}, \hat{v}$ and $\hat{T}$. In this manner Schmid & Kytömaa (1994) are able to obtain values (according to linearised theory) for the quantity

$$G(t) = \sup_{q_0 \neq 0} \frac{\|\mathbf{q}\|}{\|q_0\|}$$

where $\mathbf{q}(t) = (\hat{\nu}(t), \hat{u}(t), \hat{v}(t), \hat{T}(t))^T$, and $\mathbf{q}_0$ is the initial value of $\mathbf{q}$. The quantity $G(t)$ measures the amplification of disturbances and in some sense plays the same role as the kinetic energy in Sect. 8.1.

Many detailed results are presented in Schmid & Kytömaa (1994) and in particular, their Fig. 6 clearly displays that $G(t)$ grows relatively rapidly and may then decay or have asymptotic growth.

Wang *et al.* (1996) treats the realistic problem of shear flow of a granular medium in a *bounded* spatial layer. The question of appropriate boundary conditions is of some concern and this is analysed carefully by Wang *et al.* (1996). One important feature is that in the bounded layer case one *cannot* expect a steady shear solution to the partial differential equations (8.3.1)–(8.3.3) which is a linear function of y. Indeed, the velocity, solid fraction and granular temperature have highly non-linear profiles which have to be calculated numerically from equations (8.3.1)–(8.3.3). A selection of such profiles is displayed by Wang *et al.* (1996) in their Fig. 2. Wang *et al.* (1996) analyse the linear instability of this non-linear shear flow for a two-dimensional spatial perturbation. Their conclusions are very revealing. Among several things found by Wang *et al.* (1996) they predict the possibility of clusters of grains which will convect with the material. They argue that such clusters of short wavelength may have large transient growth even if they are asymptotically stable (according to linear theory). As far as I am aware an analysis of some appropriate measure of solution growth (such as an energy) has not been performed for the problem of shear flow in a bounded granular material. Such an analysis will be technically and computationally intensive, but should prove very rewarding.

8.4 Energy Growth in Parallel Flows of Superimposed Viscous Fluids

We conclude this book by mentioning another area where current work is ongoing to study the growth of the kinetic energy of a solution. This application is to Couette, Poiseuille and related flows in layers or pipes which contain more than one type of fluid. Flows of two or more immiscible fluids have a tremendous amount of industrial applications. For example, flow of oil and water in a pipe, flow of air over de-icer on an aeroplane wing, Chen & Crighton (1994), and there are many more. The analysis of growth of kinetic energy for one fluid overlying another in a bounded layer, say $y \in (-1, 1)$, will need computation of the spectrum of the eigenvalue problem arising from a perturbation analysis. Calculation of such spectra has already begun, Hooper & South (1995) and Dongarra *et al.* (1996), and the spectra reveal several peculiar characteristics. To my knowledge, actual calculations of kinetic energy growth have not yet been performed, although the spectra calculations of Hooper & South (1995) and Dongarra *et al.* (1996) are of interest in themselves.

For a configuration in which fluid 2 overlies fluid 1 and the fluids occupy the infinite three-dimensional region between the planes $y = -1$ and $y = n$, the basic flow takes the form, cf. Hooper & South (1995), Dongarra *et al.* (1996),

$$u_1(y) = A_1 y^2 + a_1 y + 1, \qquad -1 < y < 0,$$
$$u_2(y) = A_2 y^2 + a_2 y + 1, \qquad 0 < y < n,$$

where the A_1, A_2, a_1, a_2 are constants which depend on the viscosity ratio $m = \mu_2/\mu_1$ and on the depth ratio $n = d_2/d_1$. For a two-dimensional (spatial) perturbation the equations governing the effective perturbation stream functions become an Orr–Sommerfeld equation in each layer, namely

$$\mathcal{D}^2 \phi_1(y) = iaR\left([u_1(y) - c]\mathcal{D} - \frac{d^2 u_1}{dy^2} \right)\phi_1(y), \qquad y \in (-1, 0), \quad (8.4.1)$$

$$\mathcal{D}^2 \phi_2(y) = \frac{iaR}{m}\left([u_2(y) - c]\mathcal{D} - \frac{d^2 u_2}{dy^2} \right)\phi_2(y), \qquad y \in (0, n), \quad (8.4.2)$$

where the operator $\mathcal{D}$ is defined by

$$\mathcal{D} = \frac{d^2}{dy^2} - a^2.$$

The boundary conditions which the functions ϕ_1 and ϕ_2 must satisfy on the bounding planes are

$$\phi_1(-1) = 0, \qquad \phi_1'(-1) = 0, \qquad \phi_2(n) = 0, \qquad \phi_2'(n) = 0. \qquad (8.4.3)$$

Because one fluid overlies the other there are interface conditions which must also be satisfied. These arise because of continuity of velocity and continuity of shear and normal stresses. These boundary conditions play the mathematical role of coupling the functions ϕ_1 and ϕ_2 and are, at the interface $y = 0$,

$$\phi_1 = \phi_2,$$
$$\phi_1' = \phi_2' + \frac{\phi_1}{[c - u_1(0)]} u_1'(0)\left(\frac{1-m}{m}\right),$$
$$\phi_1'' + a^2\phi_1 = m(\phi_2'' + a^2\phi_2),$$
$$-iaR\Big([c - u_1(0)]\phi_1' + a\phi_1\Big) + iraR\Big([c - u_1(0)]\phi_2' + a_2\phi_2\Big)$$
$$-\phi_1''' + 3a^2\phi_1' + m(\phi_2''' - 3a^2\phi_2')$$
$$= \frac{iaR}{[c - u_1(0)]}\left(\frac{1}{F^2} + a^2 S\right)\phi_1. \qquad (8.4.4)$$

In these expressions $r = \rho_1/\rho_2$, with ρ_1 and ρ_2 being fluid densities, F is a non-dimensional form of the density difference, and S is a non-dimensional form of the surface tension. The quantity R is a Reynolds number and a and c are the wavenumber and growth rate which arise due to the solution form $\exp\left[ia(x - ct)\right]$.

A very convenient way to solve the eigenvalue problem (8.4.1)–(8.4.4) is to write the differential equations in (8.4.1), (8.4.2) as systems of second order equations, Dongarra $et~al.$ (1996). These writers transform the regions $y \in (-1, 0)$, $y \in (0, n)$ each to the domain $(-1, 1)$ by means of the transformations

$$z_1 = -2y - 1, \quad z_2 = \frac{2}{n}y - 1.$$

The interface then becomes $z = -1$. The functions ϕ and ψ are defined by $\phi(z_1) = \phi_1(y)$, $\psi(z_2) = \phi_2(y)$ and then the differential operators L_1 and L_2 and the functions ξ, ω are defined by

$$L_1\phi \equiv \left(4\frac{d^2}{dz_1^2} - a^2\right)\phi,$$

$$L_1\phi = \xi,$$

$$L_2\psi \equiv \left(\frac{4}{n^2}\frac{d^2}{dz_2^2} - a^2\right)\psi,$$

$$L_2\psi = \omega.$$

In this way the differential equations (8.4.1) and (8.4.2) may be arranged as the system

$$L_1\phi - \xi = 0,$$
$$L_1\xi - iaRu_1\xi + 2A_1iaR\phi = -ciaR\xi,$$
$$L_2\psi - \omega = 0,$$
$$L_2\omega - \frac{iaR}{m}u_2\omega + 2A_2i\frac{aR}{m}\psi = -ci\frac{aR}{m}\omega.$$

Dongarra $et~al.$ (1996) show how this is a natural form for application of the Chebyshev tau technique. The resulting generalised matrix eigenvalue problem is solved by using the QZ algorithm. This in itself is interesting because the eigenvalue c occurs in the boundary conditions and has to be incorporated into the matrices in a suitable way. I believe the prescription given by Dongarra $et~al.$ (1996) is a convenient way to calculate the spectra of two fluid flows in a very accurate and efficient manner. Use of such a technique should facilitate the calculation of transient kinetic energy growth in the two (or multi) layer fluid Couette and Poiseuille flow stability problems. Future results for these problems on the basis of the Butler & Farrell (1992) method outlined in Sect. 8.1 should prove revealing and will undoubtedly yield useful information on a class of flows with a multitude of industrial applications.

Bibliography

Aguirre, J. & Escobedo, M. (1993): On the blow up of solutions of a convective reaction diffusion equation. *Proc. Roy. Soc. Edinburgh A* **123**, 433–460.

Alikakos, N.D., Bates, P.W. & Grant, C.P. (1989): Blow up for a diffusion-advection equation. *Proc. Roy. Soc. Edinburgh A* **113**, 181–190.

Ames, K., Payne, L.E. & Straughan, B. (1989): On the possibility of global solutions for variant models of viscous flow on unbounded domains. *Int. J. Engng. Sci.* **27**, 755–766.

Ames, K. & Straughan, B. (1995): Boundedness and global non-existence for a non-linear convective parabolic system. *J. Math. Anal. Appl.* **190**, 795–805.

Ames, K. & Straughan, B. (1997): *Non-Standard and Improperly Posed Problems.* Mathematics in Science and Engineering Vol. 194. Academic Press, New York.

Anderson, E., Bai, Z., Bischof, C., Demmel, J., Dongarra, J.J., Du Croz, J., Greenbaum, A., Hammarling, S., McKenney, A., Ostrouchov, S. & Sorensen, D. (1995): LAPACK Users' guide, 2nd ed.

Anderson, J.R. (1991): Local existence and uniqueness of solutions of degenerate parabolic equations. *Commun. Part. Diff. Equations* **16**, 105–143.

Anderson, J.R. (1993a): Stability and instability for solutions of the convective porous medium equation with a non-linear forcing at the boundary I. *J. Differential Equations* **104**, 361–385.

Anderson, J.R. (1993b): Stability and instability for solutions of the convective porous medium equation with a non-linear forcing at the boundary II. *J. Differential Equations* **104**, 386–408.

Anderson, J.R. & Deng, K. (1995): Global existence for non-linear diffusion equations. *J. Math. Anal. Appl.* **196**, 479–501.

Antontsev, S.N. & Diaz, J.I. (1991): Space and time localization in the flow of two immiscible fluids through a porous medium: energy methods applied to systems. *Nonlin. Anal. Theory Meths. Applicns.* **16**, 299–313.

Babić, M. (1993): On the stability of rapid granular flows. *J. Fluid Mech.* **254**, 127–150.

Bair, S. & Khonsari, M. (1996): On an apparent singularity in the flow of liquids under high shear stress. ASME Symp. Rheology and Fluid Mech., Nonlinear Materials, Atlanta, paper no. G01021.

Bair, S. & Khonsari, M. (1997): A fundamental limitation of the Reynolds equation for piezoviscous liquids. Manuscript. Dept. Mech. Engng., Southern Illinois Univ.

Baker, G., Caflisch, R.E. & Siegel, M. (1993): Singularity formation during Rayleigh–Taylor instability. *J. Fluid Mech.* **252**, 51–78.

Baker, G., Meiron, D.I. & Orszag, S.A. (1982): Generalized vortex methods for free-surface flow problems. *J. Fluid Mech.* **123**, 477–501.

Ball, J.M., Holmes, P.J., James, R.D., Pego, R.L. & Swart, P.J. (1991): On the dynamics of fine structure. *J. Nonlinear Sci.* **1**, 17–70.

Bandle, C. & Levine, H.A. (1989a): On the existence and non-existence of global solutions of reaction–diffusion equations in sectorial domains. *Trans. Amer. Math.
Soc.* **655**, 595–624.

Bandle, C. & Levine, H.A. (1989b): Fujita type results for convective reaction–
diffusion equations in exterior domains. *ZAMP* **40**, 655–676.

Bandle, C. & Levine, H.A. (1994): Fujita type phenomena for reaction–diffusion
equations with convection like terms. *Diff. Integral Equations* **7**, 1169–1193.

Barron, E.N., Jensen, R. & Liu, W. (1996): Optimal control of the blowup time of
a diffusion. *Math. Models Methods Appl. Sci.* **6**, 665–687.

Bartuccelli, M., Constantin, P., Doering, C.R., Gibbon, J.D. & Gisselfält, M. (1990):
On the possibility of soft and hrad turbulence in the complex Ginzburg–Landau
equation. *Physica D* **44**, 421–444.

Beale, J.T., Kato, T. & Majda, A. (1984): Remarks on the breakdown of smooth
solutions for the 3-D Euler equations. *Commun. Math. Phys.* **94**, 61–66.

Bebernes, J. & Bricher, S. (1992): Final time blowup profiles for semilinear parabolic
equations via center manifold theory. *SIAM J. Math. Anal.* **23**, 852–869.

Bellomo, N., Brzezniak, Z. & de Socio, L.M. (1992): *Nonlinear Stochastic Problems
in Applied Sciences.* Kluwer, Dordrecht.

Bellomo, N. & Preziosi, L. (1995): *Modelling Mathematical Methods and Scientific
Computation.* CRC Press, Boca Raton.

Bellomo, N. & Riganti, R. (1987): *Nonlinear Stochastic Problems in Physics and
Mechanics.* World Scientific, Singapore.

Benjamin, T.B., Bona, J.L. & Mahony, J.J. (1972): Model equations for long waves
in non-linear, dispersive media. *Phil. Trans. Roy. Soc. London A* **272**, 47–78.

Berger, M. & Kohn, R.V. (1988): A rescaling algorithm for the numerical calculation
of blowing up solutions. *Comm. Pure Appl. Math.* **41**, 841–863.

Berlin, S., Lundbladh, A. & Henningson, D. (1994): Spatial simulations of oblique
transition in a boundary layer. *Phys. Fluids* **6**, 1949–1951.

Bona, J.L., Dougalis, V.A., Karakashian, O.A. & McKinney, W.R. (1992): Computations of blow-up and decay for periodic solutions of the generalized Korteweg–
de Vries–Burgers equation. *Appl. Numer. Math.* **10**, 335–355.

Bona, J.L., Dougalis, V.A., Karakashian, O.A. & McKinney, W.R. (1995): Conservative, high-order numerical schemes for the generalised Korteweg–de Vries
equation. *Phil. Trans. Roy. Soc. London A* **351**, 107–164.

Bona, J.L. & Saut, J.C. (1993): Dispersive blowup of solutions of generalized
Korteweg–de Vries equations. *J. Math. Anal. Appl.* **103**, 3–57.

Bricher, S. (1994): Blow-up behaviour for non-linearly perturbed semilinear para
-bolic equations. *Proc. Roy. Soc. Edinburgh A* **124**, 947–969.

Budd, C.J., Dold, J.W. & Stuart, A.M. (1994): Blow-up in a system of partial differential equations with conserved first integral. Part II. Problems with convection.
SIAM J. Appl. Math. **54**, 610–640.

Budd, C.J. & Galaktionov, V.A. (1996): Critical diffusion exponents for self-similar
blow-up solutions of a quasi-linear parabolic equation with an exponential source.
Proc. Roy. Soc. Edinburgh A **126**, 413–441.

Butler, K.M. & Farrell, B.F. (1992): Three-dimensional optimal perturbations in
viscous shear flow. *Phys. Fluids A* **4**, 1637–1650.

Butler, K.M. & Farrell, B.F. (1993): Optimal perturbations and streak spacing in
wall-bounded turbulent shear flow. *Phys. Fluids A* **5**, 774–777.

Butler, K.M. & Farrell, B.F. (1994): Nonlinear equilibration of two-dimensional
optimal perturbations in viscous shear flow. *Phys. Fluids A* **6**, 2011–2020.

Caflisch, R.E. & Orellana, O.F. (1989): Singular solutions and ill-posedness for the
evolution of vortex sheets. *SIAM J. Math. Anal.* **20**, 293–307.

Canuto, C., Hussaini, M.Y., Quarteroni, A. & Zang, T.A. (1988): *Spectral Methods in Fluid Dynamics*. Springer, Berlin, Heidelberg.

Chadam, J.M., Peirce, A. & Yin, H.M. (1992): The blow-up property of solutions to some diffusion equations with localized non-linear reactions. *J. Math. Anal. Appl.* **169**, 313–328.

Chadam, J.M. & Yin, H.M. (1993): A diffusion equation with localized chemical reactions. *Proc. Edinburgh Math. Soc.* **37**, 101–118.

Chen, K.P. & Crighton, D.G. (1994): Instability of the large Reynolds number flow of a Newtonian fluid over a viscoelastic fluid. *Phys. Fluids A* **6**, 152–163.

Chen, P.J. (1973): Growth and decay of waves in solids. In, *Handbuch der Physik* VIa/3. Ed. C. Truesdell, Springer, Berlin, Heidelberg.

Childress, S., Ierley, G.R., Spiegel, E.A. & Young, W.R. (1989): Blow up of unsteady two-dimensional Euler and Navier–Stokes solutions having stagnation-point form. *J. Fluid Mech.* **203**, 1–22.

Chi-Hwa Wang, Jackson, R. & Sunderesan, S. (1996): Stability of bounded rapid shear flows of a granular material. *J. Fluid Mech.* **308**, 31–62.

Chipot, M. & Weissler, F.B. (1989): Some blow up results for a non-linear parabolic equation with gradient term. *SIAM J. Math. Anal.* **20**, 886–907.

Constantin, P., Lax, P. & Majda, A. (1985): A simple one-dimensional model for the three-dimensional vorticity equation. *Comm. Pure Appl. Math.* **38**, 715–724.

Constantin, P., Majda, A. & Tabak, E. (1994): Singular front propagation in a model for quasigeostrophic flow. *Phys. Fluids* **6**, 9–11.

Cox, E.A. & Mortell, M.P. (1983): The evolution of resonant oscillations in closed tubes. *ZAMP* **34**, 845–866.

Cox, S.M. (1991): Two-dimensional flow of a viscous fluid in a channel with porous walls. *J. Fluid Mech.* **227**, 1–33.

Dafermos, C.M. (1985): Contemporary issues in the dynamic behavior of continuous media. LCDS Lecture Notes, 85-1, Brown University.

Dafermos, C.M. (1986): Development of singularities in the motion of materials with fading memory. *Arch. Rational Mech. Anal.* **91**, 193–205.

Dafermos, C.M. & Hsaio, L. (1986): Development of singularities in solutions of the equations of non-linear thermoelasticity. *Quart. Appl. Math.* **64**, 463–474.

Davis, S.H. (1969): On the principle of exchange of stabilities. *Proc. Roy. Soc. London A* **310**, 341–358.

De Gregorio, S. (1990): On a one-dimensional model for the three-dimensional vorticity equation. *J. Statistical Phys.* **59**, 1251–1263.

DelSole, T. & Farrell, B.F. (1995): A stochastically excited linear system as a model for quasigeostrophic turbulence: analytic results for one- and two-layer fluids. *J. Atmos. Sci.* **52**, 2531–2547.

DelSole, T. & Farrell, B.F. (1996): The quasi-linear equilibration of a thermally maintained, stochastically excited jet in a quasigeostrophic model. *J. Atmos. Sci.* **53**, 1781–1797.

Deng, K. (1994): Behavior of solutions of Burgers' equation with non-local boundary conditions. II. *Quart. Appl. Math.* **52**, 553–567.

Deng, K. (1995): Global existence and blow-up for a system of heat equations with non-linear boundary conditions. *Math. Meth. Appl. Sci.* **18**, 307–315.

Deng, K., Kwong, M.K. & Levine, H.A. (1992): The influence of non-local non-linearities on the long-time behaviour of solutions of Burgers' equation. *Quart. Appl. Math.* **50**, 173–200.

Desai, C.S., Samtani, N.C. & Vulliet, L. (1995): Constitutive modelling and analysis of creeping slopes. *J. Geotech. Engng.* **121**, 43–56.

Di Benedetto, E. & Friedman, A. (1984): The ill-posed Hele–Shaw model and the Stefan problem for supercooled water. *Trans. Amer. Math. Soc.* **282**, 183–204.

Dlotko, T. (1991): Examples of parabolic problems with blowing-up derivatives. *J. Math. Anal. Appl.* **154**, 226–237.

Doering, C.R. & Gibbon, J.D. (1995): *Applied Analysis of the Navier–Stokes Equations*. Cambridge University Press, Cambridge.

Dongarra, J.J., Straughan, B. & Walker, D.W. (1996): Chebyshev tau–QZ algorithm methods for calculating spectra of hydrodynamic stability problems. *Appl. Numer. Math.* **22**, 399–435.

Drazin, P.G. (1991): Stability of non-parallel flow in a channel. *Le Matematiche* **46**, 137–146.

Dunn, J.E. & Fosdick, R.L. (1974): Thermodynamics, stability and boundedness of fluids of complexity 2 and fluids of second grade. *Arch. Rational Mech. Anal.* **56**, 191–252.

Escobedo, M. & Herrero, M.A. (1991): Boundedness and blow up for a semilinear reaction–diffusion system. *J. Differential Equations* **89**, 176–202.

Escobedo, M. & Herrero, M.A. (1993): A semilinear parabolic system in a bounded domain. *Ann. Matem. Pura Appl.* **165**, 315–336.

Escobedo, M. & Levine, H.A. (1995): Critical blow up and global existence numbers for a weakly coupled system of reaction–diffusion equations. *Arch. Rational Mech. Anal.* **129**, 47–100.

Etheridge, A.M. (1996): A probabilistic approach to blow-up of a semilinear heat equation. *Proc. Roy. Soc. Edinburgh A* **126**, 1235–1245.

Farrell, B.F. (1988a): Optimal excitation of perturbations in viscous shear flow. *Phys. Fluids* **31**, 2093–2102.

Farrell, B.F. (1988b): Optimal excitation of neutral Rossby waves. *J. Atmos. Sci.* **45**, 163–172.

Farrell, B.F. (1989): Optimal excitation of baroclinic waves. *J. Atmos. Sci.* **46**, 1193–1206.

Farrell, B.F. (1990): Small error dynamics and the predictability of atmospheric flows. *J. Atmos. Sci.* **47**, 2409–2416.

Farrell, B.F. & Ioannou, P.J. (1993a): Optimal excitation of three-dimensional perturbations in viscous constant shear flow. *Phys. Fluids A* **5**, 1390–1400.

Farrell, B.F. & Ioannou, P.J. (1993b): Stochastic forcing of the linearized Navier–Stokes equations. *Phys. Fluids A* **5**, 2600–2609.

Farrell, B.F. & Ioannou, P.J. (1993c): Stochastic forcing of perturbation variance in unbounded shear and deformation flows. *J. Atmos. Sci.* **50**, 200–211.

Farrell, B.F. & Ioannou, P.J. (1993d): Stochastic dynamics of baroclinic waves. *J. Atmos. Sci.* **50**, 4044–4057.

Farrell, B.F. & Ioannou, P.J. (1994a): Variance maintained by stochastic forcing of non-normal dynamical systems associated with linearly stable shear flows. *Phys. Rev. Letters* **72**, 1188–1191.

Farrell, B.F. & Ioannou, P.J. (1994b): A theory for the statistical equilibrium energy spectrum and heat flux produced by transient baroclinic waves. *J. Atmos. Sci.* **51**, 2685–2698.

Farrell, B.F. & Ioannou, P.J. (1995): Stochastic dynamics of the midlatitude atmospheric jet. *J. Atmos. Sci.* **52**, 1642–1656.

Farrell, B.F. & Ioannou, P.J. (1996a): Turbulence suppression by active control. *Phys. Fluids* **8**, 1257–1268.

Farrell, B.F. & Ioannou, P.J. (1996b): Generalised stability theory. Part I: autonomous operators. *J. Atmos. Sci.* **53**, 2025–2040.

Farrell, B.F. & Ioannou, P.J. (1996c): Generalised stability theory. Part II: non-autonomous operators. *J. Atmos. Sci.* **53**, 2041–2053.

Farrell, B.F. & Ioannou, P.J. (1998): Perturbation structure and spectra in turbulent channel flow. *J. Comp. Fluid Dyn.*, in the press.

Ferrari, A.B. (1993): On the blow-up of solutions of the 3-D Euler equations in a bounded domain. *Comm. Math. Phys.* **155**, 277–294.

Fila, M. (1991): Remarks on blow up for a non-linear parabolic equation with a gradient term. *Proc. Amer. Math. Soc.* **111**, 795–801.

Fila, M., Levine, H.A. & Uda, Y. (1994): A Fujita-type global existence–global non-existence theorem for a system of reaction diffusion equations with differing diffusivities. *Math. Meth. Appl. Sci.* **17**, 807–835.

Fila, M. & Lieberman, G.M. (1994): Derivative blow-up and beyond for quasilinear parabolic equations. *Diff. Integral Equations* **7**, 811–821.

Filippas, S. & Kohn, R.V. (1992): Refined asymptotics for the blowup of $u_t - \Delta u = u^p$. *Comm. Pure Appl. Math.* **45**, 821–869.

Flavin, J.N. & Rionero, S. (1995): *Qualitative Estimates for Partial Differential Equations*. CRC Press, Boca Raton.

Floater, M.S. (1991): Blow-up at the boundary for degenerate semilinear parabolic equations. *Arch. Rational Mech. Anal.* **114**, 57–77.

Foda, M.A. (1987): Internal dissipative waves in poroelastic media. *Proc. Roy. Soc. London A* **413**, 383–405.

Foda, M.A. (1994): Landslides riding on basal pressure waves. *Continuum Mech. Thermodyn.* **6**, 61–79.

Fosdick, R.L. & Rajagopal, K.R. (1979): Anomalous features in the model of second order fluids. *Arch. Rational Mech. Anal.* **70**, 145–152.

Fosdick, R.L. & Rajagopal, K.R. (1980): Thermodynamics and stability of fluids of third grade. *Proc. Roy. Soc. London A* **339**, 351–377.

Fosdick, R.L. & Straughan, B. (1981): Catastrophic instabilities and related results in a fluid of third grade. *Int. J. Nonlinear Mech.* **16**, 191–198.

Franchi, F. & Straughan, B. (1993): Stability and non-existence results in the generalized theory of a fluid of second grade. *J. Math. Anal. Appl.* **180**, 122–137.

Friedman, A. (1988): Blow-up of solutions of non-linear parabolic equations. In *Non-linear Diffusion Equations and their Equilibrium States I.* Eds. W.M. Ni, L.A. Peletier & J. Serrin. Springer, Berlin, Heidelberg. pp. 301–318.

Friedman, A. & Giga, Y. (1987): A single point blow-up for positive solutions of semilinear parabolic systems. *Fac. Sci. Univ. Tokyo, Sect. IA* **34**, 65–79.

Friedman, A. & Lacey, A.A. (1988): Blow-up of solutions of semilinear parabolic equations. *J. Math. Anal. Appl.* **132**, 171–186.

Friedman, A. & McLeod, J.B. (1985): Blow-up of positive solutions of semilinear heat equations. *Indiana Univ. Math. J.* **34**, 425–447.

Fu, Y.B. & Scott, N.H. (1991): The transistion from acceleration wave to shock wave. *Int. J. Engng. Sci.* **29**, 617–624.

Fujita, H. (1966): On the blowing up of solutions of the Cauchy problem for $u_t = \Delta u + u^{1+\alpha}$. *J. Fac. Sci. Univ. Tokyo, Sect. I* **13**, 109–124.

Fujita, H. (1970): On some non-existence and non-uniqueness theorems for non-linear parabolic equations. *Proc. Symp. Pure Math.* **18**, 105–113.

Giga, Y. & Kohn, R.V. (1985): Asymptotically self-similar blow up of semilinear heat equations. *Comm. Pure Appl. Math.* **38**, 297–319.

Giga, Y. & Kohn, R.V. (1987): Characterizing blow up using similarity variables. *Indiana Univ. Math. J.* **36**, 1–39.

Glenn Lasseigne, D. & Olmstead, W.E. (1983): Ignition of a combustible solid by convection heating. *ZAMP* **34**, 886–898.

Glenn Lasseigne, D. & Olmstead, W.E. (1991): Ignition or non-ignition of a combustible solid with marginal heating. *Quart. Appl. Math.* **49**, 309–312.

Goldshtein, V., Zinoviev, A., Sobolev, V. & Shchepakina, E. (1996): Criterion for thermal explosion with reactant consumption in a dusty gas. *Proc. Roy. Soc. London A* **452**, 2103–2119.

Golovkin, K.K. (1967): New model equations of motion of a viscous fluid and their unique solvability. *Trudy Mat. Inst. Steklov* **102**, 29–50.

Gray, J.M.N.T. & Killworth, P.D. (1995): Stability of the viscous-plastic sea ice rheology. *J.Phys. Oceanography* **25**,971–978.

Gupta, G. & Massoudi, M. (1993): Flow of a generalized second grade fluid between heated plates. *Acta Mech.* **99**, 21–33.

Gustavsson, L.H. (1981): Resonant growth of three-dimensional disturbances in plane Poiseuille flow. *J. Fluid Mech.* **112**, 253–264.

Gustavsson, L.H. (1986): Excitation of direct resonances in plane Poiseuille flow. *Stud. Appl. Math.* **75**, 227–248.

Gustavsson, L.H. (1991): Energy growth of three-dimensional disturbances in plane Poiseuille flow. *J. Fluid Mech.* **224**, 241–260.

Gustavsson, L.H. & Hultgren, L.S. (1980): A resonance mechanism in plane Couette flow. *J. Fluid Mech.* **98**, 149–159.

Hadamard, J. (1922): *Lectures on Cauchy's Problem.* Yale University Press, New Haven, CT.

Hall, P., Balakumar, P. & Papageorgiu, D. (1992): On a class of unsteady three-dimensional Navier–Stokes solutions relevant to rotating disc flows: threshold amplitudes and finite-time singularities. *J. Fluid Mech.* **238**, 297–323.

Hanifi, A., Schmid, P.J. & Henningson, D.S. (1996): Transient growth in compressible boundary layer flow. *Phys. Fluids* **8**, 1–12.

Hayakawa, (1973): On non-existence of global solutions of some semi-linear parabolic differential equations. *Proc. Japan Acad.* **49**, 503–505.

Henningson, D.S. (1995): Bypass transition and linear growth mechanisms. In *Advances in Turbulence V.* Ed. R. Benzi. Kluwer, Dordrecht. pp. 190–204.

Henningson, D.S., Lundbladh, A. & Johansson, A.V. (1993): A mechanism for bypass transition from localized disturbances in wall-bounded shear flows. *J. Fluid Mech.* **250**, 169–207.

Henningson, D.S. & Reddy, S.C. (1994): On the role of linear mechanisms in transition to turbulence. *Phys. Fluids* **6**, 1396–1398.

Hill, R. (1958): A general theory of uniqueness and stability in elastic-plastic solids. *J. Mech. Phys. Solids* **6**, 236–249.

Hooper, A.P. & Grimshaw, R. (1995): Disturbance growth of linearly stable viscous shear flows. Research and Consultancy Report 3. Dept. Math. Sciences, Univ. West of England.

Hooper, A.P. & South, M. (1995): Eigenvalues and disturbance growth in channel flows of two superposed viscous fluids. Research and Consultancy Report 4. Dept. Math. Sciences, Univ. West of England.

Hoskins, B.J. & Bretherton, F.P. (1972): Atmospheric frontogenesis models: mathematical formulation and solution. *J. Atmos. Sci.* **29**, 11–37.

Howes, F.A. (1986a): Some stability results for advection–diffusion equations. *Stud. Appl. Math.* **74**, 35–53.

Howes, F.A. (1986b): Multi-dimensional initial boundary value problems with strong non-linearities. *Arch. Rational Mech. Anal.* **91**, 153–168.

Hutter, K. (1983): *Theoretical Glaciology.* D. Reidel, Dordrecht.

Hutter, K., Szidarovszky, F. & Yakowitz, S. (1987): Granular shear flows as models for flow avalanches. Results of a numerical study. In *International Symposium on Avalanche Formation, Movement and Effects.* Davos, Switzerland. International Association of Hydrological Sciences. Vol. 162. pp. 381–394.

Jacqmin, D. (1991): Frontogenesis driven by horizontally quadratic distributions of density. *J. Fluid Mech.* **228**, 1–24.

Jäger, W. & Luckhaus, S. (1992): On explosions of solutions to a system of partial differential equations modelling chemotaxis. *Trans. Amer. Math. Soc.* **329**, 819–824.

Jones, D.R. (1973): The dynamic stability of confined, exothermically reacting fluids. *Int. J. Heat Mass Transfer* **16**, 157–167.

Jones, D.R. (1974): Convective effects in enclosed, exothermically reacting gases. *Int. J. Heat Mass Transfer* **17**, 11–21.

Joseph, D.D. (1976): *Stability of Fluid Motions I.* Springer, Berlin, Heidelberg.

Joseph, D.D., Renardy, M. & Saut, J.C. (1985): Hyperbolicity and change of type in the flow of viscoelastic fluids. *Arch. Rational Mech. Anal.* **87**, 213–251.

Joseph, D.D. & Saut, J.C. (1990): Short wave instabilities and ill posed initial value problems. *Theoret. Comput. Fluid Dyn.* **1**, 191–227.

Junning, Z. (1993): Existence and non-existence of solutions for $u_t = \operatorname{div}(|\nabla u|^{p-2}\nabla u) + f(\nabla u, u, x, t)$. *J. Math. Anal. Appl.* **172**, 130–146.

Kalantarov, V.K. & Ladyzhenskaya, O.A. (1977): On the origin of collapses for quasilinear equations of parabolic and hyperbolic type. In *Boundary Problems of Mathematical Physics and Related Questions of the Function Theory, 10.* Ed. O.A. Ladyzhenskaya. Zapiski, Nauchnih Seminarov LOMI, Vol. 69. Nauka, Moscow. pp. 77–102.

Kawohl, B. & Peletier, L.A. (1989): Observations on blow up and dead cores for non-linear parabolic equations. *Math. Zeitschrift* **202**, 207–217.

Kapila, A.K. (1981): Evolution of deflagration in a cold combustible subjected to a uniform energy flux. *Int. J. Engng. Sci.* **19**, 495–509.

Keller, J.B. (1957): On the solutions of non-linear wave equations. *Comm. Pure Appl. Math.* **10**, 523–530.

Keller, E.F. & Segel, L.A. (1970): Initiation of slime mold aggregation viewed as an instability. *J. Theor. Biol.* **26**, 399–415.

Keller, E.F. & Segel, L.A. (1971a): Model for chemotaxis. *J. Theor. Biol.* **30**, 225–234.

Keller, E.F. & Segel, L.A. (1971b): Traveling bands of chemotactic bacteria: a theoretical analysis. *J. Theor. Biol.* **30**, 235–248.

Kerswell, R.R. (1997): Variational bounds on shear driven turbulence and turbulent Boussinesq convection. *Physica D* **100**, 355–376.

Kerswell, R.R. & Davey, A. (1996): On the linear instability of elliptic pipe flow. *J. Fluid Mech.* **316**, 307–324.

Kreiss, G., Lundbladh, A. & Henningson, D.S. (1994): Bounds for threshold amplitudes in subcritical shear flows. *J. Fluid Mech.* **270**, 175–198.

Kutev, N. (1991): On the solvability of Dirichlet's problem for a class of non-linear elliptic and parabolic equations. In *Proc. Equadiff 1991,* vol. 2. Eds. C. Perello *et al.* World Scientific, Singapore, pp. 666–670.

Kutev, N. (1992): Global solvability and boundary gradient blow up for one dimensional parabolic equations. In *Progress in Partial Differential Equations, Elliptic and Parabolic Problems.* Eds. C. Bandle *et al.* Longman, New York. pp. 176–181.

Kutev, N. (1994): Gradient blow ups and global solvability after the blow up time for non-linear parabolic equations. In *Proc. 3rd Intl. Conf. Evolution Equations, Control Theory and Biomathematics.* Eds. P. Clement and G. Lumer. Marcel Dekker, New York. pp. 301–306.

Ladyzhenskaya, O.A. (1967): New equations for the description of motions of viscous incompressible fluids and global solvability of their boundary value problems. *Trudy Mat. Inst. Steklov* **102**, 85–104.

Ladyzhenskaya, O.A. (1968): On some non-linear problems in the theory of continuous media. *Amer. Math. Soc. Translations* **70**, 73–89.

Ladyzhenskaya, O.A. (1969): *The Mathematical Theory of Viscous Incompressible Flow*. 2nd ed. Gordon and Breach, New York.

Leray, J. (1934): Sur le mouvement d'un liquide visqueux emplissant l'espace. *Acta Math*. **63**, 193–248.

Levine, H.A. (1972): Some uniqueness and growth theorems in the Cauchy problem for $Pu_{tt} + Mu_t + Nu = 0$ in Hilbert space. *Math. Zeitschrift* **126**, 345–360.

Levine, H.A. (1973): Some non-existence and instability theorems for solutions of formally parabolic equations of the form $Pu_t = -Au + \mathcal{F}(u)$. *Arch. Rational Mech. Anal*. **51**, 371–386.

Levine, H.A. (1974): On the non-existence of global solutions to a non-linear Euler–Poisson–Darboux equation. *J. Differential Equations* **48**, 646–651.

Levine, H.A. (1975): Non-existence of global weak solutions to some properly and improperly posed problems of mathematical physics: the method of unbounded Fourier coefficients. *Math. Annalen* **214**, 205–220.

Levine, H.A. (1988): Stability and instability for solutions of Burgers' equation with a semilinear boundary condition. *SIAM J. Math. Anal*. **19**, 312–336.

Levine, H.A. (1990): The role of critical exponents in blowup theorems. *SIAM Review* **32**, 262–288.

Levine, H.A. (1991): A Fujita type global existence – global non-existence theorem for a weakly coupled system of reaction–diffusion equations. *ZAMP* **42**, 408–430.

Levine, H.A., Lieberman, G.M. & Meier, P. (1990): On critical exponents for some quasilinear parabolic equations. *Math. Methods Appl. Sci*. **12**, 429–438.

Levine, H.A. & Meier, P. (1989): A blow-up result for the critical exponents in cones. *Israel J. Math*. **67**, 1–7.

Levine, H.A. & Meier, P. (1990): The value of the critical exponent for reaction–diffusion equations in cones. *Arch. Rational Mech. Anal*. **109**, 73–80.

Levine, H.A. & Payne, L.E. (1974a): Non-existence theorems for the heat equation with non-linear boundary conditions and for the porous medium equation backward in time. *J. Differential Equations* **16**, 319–334.

Levine, H.A. & Payne, L.E. (1974b): Some non-existence theorems for initial boundary value problems with non-linear boundary constraints. *Proc. Amer. Math. Soc*. **46**, 277–284.

Levine, H.A., Payne, L.E., Sacks, P.E. & Straughan, B. (1989): Analysis of a convective reaction–diffusion equation. *SIAM J. Math. Anal*. **20**, 133–147.

Levine, H.A. & Protter, M.H. (1986): The breakdown of solutions of quasilinear first order systems of partial differential equations. *Arch. Rational Mech. Anal*. **95**, 253–267.

Levine, H.A. & Sacks, P.E. (1984): Some existence and non-existence theorems for solutions of degenerate parabolic equations. *J. Differential Equations* **52**, 135–161.

Levine, H.A. & Sleeman, B.D. (1995): A system of reaction diffusion equations arising in the theory of reinforced random walks. *IMA Preprint Ser., University of Minnesota*, **1371**.

Levine, H.A. & Smith, R.A. (1986): A potential well theory for the wave equation with a non-linear boundary condition. *J. Reine Angew. Math*. **374**, 1–23.

Levine, H.A. & Smith, R.A. (1987): A potential well theory for the heat equation with a non-linear boundary condition. *Math. Methods Appl. Sci*. **9**, 127–136.

Lianjun An (1994): The genericity of flutter ill-posedness in three-dimensional elastic-plastic models. *Quart. Appl. Math*. **52**, 343–362.

Lianjun An & Peirce, A. (1994): The effect of microstructure on elastic-plastic models. *SIAM J. Appl. Math*. **54**, 708–730.

Lianjun An & Peirce, A. (1995): A weakly non-linear analysis of elasto-plastic-microstructure models. *SIAM J. Appl. Math*. **55**, 136–155.

Lianjun An & Schaeffer, D.G. (1992): The flutter instability in granular flow. *J. Mech. Phys. Solids* **40**, 683–698.

Lieberman, G.M. (1986): The first initial boundary value problem for quasilinear second order parabolic equations. *Ann. Scuola Norm. Sup. Pisa* **13**, 347–387.

Lin, C.C. & Segel, L.A. (1974): *Mathematics Applied to Deterministic Problems in the Natural Sciences.* Macmillan, New York.

Logan, J.D. (1987): *Applied Mathematics: A Contemporary Approach.* Wiley, New York.

Logan, J.D. (1994): *An Introduction to Nonlinear Partial Differential Equations.* Wiley, New York.

Lu, G. & Sleeman, B.D. (1993a): Non-existence of global solutions to systems of semi-linear parabolic equations. *J. Differential Equations* **104**, 147–168.

Lu, G. & Sleeman, B.D. (1994): Subsolutions and super solutions to systems of parabolic equations with applications to generalized Fujita-type systems. *Math. Meth. Appl. Sci.* **17**, 1005–1016.

Mack, L.M. (1976): A numerical study of the temporal eigenvalue spectrum of the Blasius boundary layer. *J. Fluid Mech.* **73**, 497–520.

Majda, A. (1986): Vorticity and the mathematical theory of incompressible fluid flow. *Comm. Pure Appl. Math.* **39**, S187–S220.

Malik, M.R. (1990): Numerical methods for hypersonic boundary layer stability. *J. Computational Phys.* **86**, 376–413.

Malik, S.K. & Singh, M. (1992): An explosive instability in magnetic fluids. *Quart. Appl. Math.* **50**, 613–626.

Man, C.S. (1992): Nonsteady channel flow of ice as a modified second-order fluid with power-law viscosity. *Arch. Rational Mech. Anal.* **119**, 35–57.

Man, C.S. & Sun, Q.K. (1987): On the significance of normal stress effects in the flow of glaciers. *J. Glaciology* **33**, 268–273.

Meier, P. (1990): On the critical exponent for reaction–diffusion equations. *Arch. Rational Mech. Anal.* **109**, 63–71.

Moler, C.B. & Stewart, G.W. (1973): An algorithm for generalized matrix eigenproblems. *SIAM. J. Numerical Anal.* **10**, 241–256.

Moore, A.M. & Farrell, B.F. (1994): Using adjoint models for stability and predictability analysis. NATO ASI Series. Vol. 119. Springer, Berlin, Heidelberg. pp. 217–239

Moore, D.W. (1979): The spontaneous appearance of a singularity in the shape of an evolving vortex sheet. *Proc. Roy. Soc. London A* **365**, 105–119.

Morro, A. (1978): Interaction of acoustic waves with shock waves in elastic solids. *ZAMP* **29**, 822–827.

Morro, A. & Straughan, B. (1983): Highly unstable solutions to completely non-linear diffusion problems. *Nonlinear Anal. Theory, Meths., Applicns.* **7**, 231–237.

Mueller, C.E. & Weissler, F.B. (1985): Single point blow-up for a general semilinear heat equation. *Indiana Univ. Math. J.* **34**, 881–913.

Mulone, G. & Rionero, S. (1993): On the non-linear stability of the magnetic Bénard problem with rotation. *Z. angew. Math. Mech.* **73**, 35–45.

Mulone, G., Rionero, S. & Straughan, B. (1996): Unconditional non-linear stability in a polarized dielectric liquid. *Atti Accad. Naz. Lincei* (Ser. IX) **7**, 241–252.

Necas, J., Ruzicka, M. & Sverak, V. (1996): On Leray's self-similar solutions of the Navier–Stokes equations. *Acta Math.* **176**, 283–294.

Neitzel, G.P., Smith, M.K. & Bolander, M.J. (1994): Thermal instability with radiation by the method of energy. *Int. J. Heat Mass Transfer* **37**, 2909–2915.

Ni, W.M., Sacks, P.E. & Tavantzis. (1984): On the asymptotic behavior of solutions of certain quasilinear parabolic equations. *J. Differential Equations* **54**, 97–120.

Nield, D.A. & Bejan, A. (1992): *Convection in Porous Media.* Springer, Berlin, Heidelberg.

Novikov, V. (1990): The internal dynamics of flows and formation of singularities. *Fluid Dyn. Res.* **6**, 79–89.

Olmstead, W.E. (1983): Ignition of a combustible half space. *SIAM J. Appl. Math.* **43**, 1–15.

Olmstead, W.E., Nemat-Nasser, S. & Ni, L. (1994): Shear bands as surfaces of discontinuity. *J. Mech. Phys. Solids* **42**, 697–709.

Olmstead, W.E. & Roberts, C.A. (1994): Explosion in a diffusive strip due to a concentrated non-linear source. *Meth. Applicns. Anal.* **1**, 434–445.

Olmstead, W.E. & Roberts, C.A. (1996): Explosion in a diffusive strip due to a source with local and non-local features. *Meth. Applicns. Anal.*, in the press.

Olmstead, W.E., Roberts, C.A. & Deng, K. (1995): Coupled Volterra equations with blow-up solutions. *J. Integral Equations & Applicns.* **7**, 499–516.

Orszag, S.A. (1971): Accurate solution of the Orr–Sommerfeld stability equation. *J. Fluid Mech.* **50**, 689–703.

Osinov, V.A. (1997): Plane waves and dynamic ill-posedness in granular media. In *Powders and Grains '97.* Eds. Behringer, R.P. and Jenkins, J.T. Balkema, Rotterdam, pp. 363-366.

Osinov, V.A. (1998): Theoretical investigation of large amplitude waves in granular soils. *Soil Dynamics and Earthquake Engineering,* **17**, 13–28.

Osinov, V.A. & Gudehus, G. (1996): Plane shear waves and loss of stability in a saturated granular body. *Mechanics for Cohesive Frictional Materials,* **1**, 25–44.

Payne, L.E. (1971): Uniqueness and continuous dependence criteria for the Navier–Stokes equations. *Rocky Mtn. J. Math.* **2**, 641–660.

Payne, L.E. (1975): *Improperly Posed Problems in Partial Differential Equations.* Regional Conf. Ser. Appl. Math., SIAM.

Payne, L.E. (1993): On stabilizing ill-posed Cauchy problems for the Navier–Stokes equations. In *Differential Equations with Applications to Mathematical Physics.* Academic Press, New York. pp. 261–271.

Payne, L.E. & Straughan, B. (1987): Unconditional non-linear stability in penetrative convection. *Geophys. Astrophys. Fluid Dyn.* **39**, 57–63. (Also, Corrected and extended numerical results. *Geophys. Astrophys. Fluid Dyn.* (1988) **43**, 307–309.)

Preziosi, L. (1993): A source-sink inverse problem for the non-linear heat equation. *Comp. Math. Model.* **17**, 3–11.

Preziosi, L. & de Socio, L. (1991): Non-linear inverse phase transistion problems for the non-linear heat equation. *Math. Models Methods Appl. Sci.* **1**, 167–182.

Preziosi, L., Teppati, G. & Bellomo, N. (1992): Modelling and solution of stochastic inverse problems in mathematical physics. *Comp. Math. Model.* **16**, 37–51.

Proudman, I. & Johnson, K. (1962): Boundary layer growth at a rear stagnation point. *J. Fluid Mech.* **12**, 161–168.

Rasmussen, C.H. (1979): Oscillation and asymptotic behavior of systems of ordinary differential equations. *Trans. Amer. Math. Soc.* **256**, 1–47.

Reddy, S.C. & Henningson, D.S. (1993): Energy growth in viscous channel flows. *J. Fluid Mech.* **252**, 209–238.

Reddy, S.C., Schmid, P.J. & Henningson, D.S. (1993): Pseudospectra of the Orr–Sommerfeld operator. *SIAM J. Appl. Math.* **53**, 15–47.

Renardy, M. (1986): Some remarks on the Navier–Stokes equations with a pressure-dependent viscosity. *Comm. Part. Diff. Equations* **11**, 779–793.

Richardson, S. (1972): Hele–Shaw flows with a free boundary produced by the injection of fluid into a narrow channel. *J. Fluid Mech.* **56**, 609–618.

Rionero, S. & Mulone, G. (1991): On the non-linear stability of parallel shear flows. *Continuum Mech. Thermodyn.* **3**, 1–11.

Roberts, C.A., Glenn Lasseigne, D. & Olmstead, W.E. (1993): Volterra equations which model explosion in a diffusive medium. *J. Integral Equations & Applicns.* **5**, 531–546.

Roberts, C.A. & Olmstead, W.E. (1996): Growth rates for blow-up solutions of non-linear Volterra equations. *Quart. Appl. Math.* **54**, 153–159.

Rosensweig, R.E. (1985): *Ferrohydrodynamics*. Cambridge University Press, Cambridge.

Samarskii, A.A., Galaktionov, V.A., Kurdyumov, S.P. & Mikhailov, A.P. (1994): *Blow-up in Quasilinear Parabolic Equations*. Walter de Gruyter, Berlin.

Samtani, N.C., Desai, C.S. & Vulliet, L. (1994): A viscoplastic model for creeping natural slopes. In *Computer Methods and Advances in Geomechanics*, Eds. Sinwardane & Zaman. Balkema, Rotterdam. pp. 2483–2488.

Savage, S.B. (1992): Instability of unbounded uniform granular shear flow. *J. Fluid Mech.* **241**, 109–123.

Schaeffer, D.G. (1990): Instability and ill-posedness in the deformation of granular materials. *Int. J. Numer. Anal. Meth. Geomech.* **14**, 253–278.

Schaeffer, D.G. (1992): A mathematical model for localization in granular flow. *Proc. Roy. Soc. London A* **436**, 217–250.

Schaeffer, D.G. & Shearer, M. (1997): The influence of material non-uniformity preceding shear-band formation in a model for granular flow. *Euro. J. Appl. Math.* **8**, 457–483.

Schmid, P.J. & Henningson, D.S. (1994): Optimal energy density growth in Hagen–Poiseuille flow. *J. Fluid Mech.* **277**, 197–225.

Schmid, P.J. & Kytömaa, H.K. (1994): Transient and asymptotic stability of granular shear flow. *J. Fluid Mech.* **264**, 255–275.

Schubert, G. & Yuen, D.A. (1982): Initiation of ice ages by creep instability and surging of the East Antarctic ice sheet. *Nature* **296**, 127–130.

Shanthini, R. (1989): Degeneracies of the temporal Orr–Sommerfeld eigenmodes in plane Poiseuille flow. *J. Fluid Mech.* **201**, 13–34.

Shearer, M. & Schaeffer, D.G. (1994): Unloading near a shear band in granular material. *Quart. Appl. Math.* **52**, 579–600.

Short, M. (1995): The initiation of detonation from general non-uniformly distributed initial conditions.. *Phil. Trans. Roy. Soc. London A* **353**, 173–203.

Short, M. (1996): Homogeneous thermal explosion in a compressible atmosphere. *Proc. Roy. Soc. London A* **452**, 1127–1138.

Smith, F.T. & Bowles, R.I. (1992): Transition theory and experimental comparisons on (I) amplification into streets and (II) a strongly non-linear break-up criterion. *Proc. Roy. Soc. London A* **439**, 163–175.

Sonneveld, P. (1989): CGS, a fast Lanczos-type solver for non-symmetric linear systems. *SIAM J. Sci. Stat. Comput.* **10**, 36-52.

Stewart, K. & Geveci, T. (1992): Numerical experiments with a non-linear evolution equation which exhibits blow-up. *Appl. Numer. Math.* **10**, 139–147.

Stewart, P.A. & Smith, F.T. (1992): Three-dimensional non-linear blow-up from a nearly planar intial disturbance, in boundary layer transistion: theory and experimental comparisions. *J. Fluid Mech.* **244**, 79–100.

Stewart, G.W. & Sun, J.-G. (1990): *Matrix Perturbation Theory*. Academic Press, New York.

Straughan, B. (1976): Global non-existence of solutions to Ladyzhenskaya's variants of the Navier-Stokes equations backward in time. *Proc. Roy. Soc. Edinburgh A* **75**, 165–170.

Straughan, B. (1982): *Instability, non-existence and weighted energy methods in fluid dynamics and related theories*. Research Notes in Mathematics, Vol. 74. Pitman, London.

Straughan, B. (1987): The backward in time problem for a fluid of third grade. *Ricerche Matem.* **36**, 289–293.

Straughan, B. (1988): A non-linear analysis of convection in a vertical porous slab. *Geophys. Astrophys. Fluid Dyn.* **42**, 269–275.

Straughan, B. (1992): *The Energy Method, Stability, and Non-linear Convection.* Series in Applied Mathematical Sciences. Vol. 91. Springer, Berlin, Heidelberg.

Straughan, B. (1993): *Mathematical aspects of penetrative convection.* Research Notes in Mathematics, Vol. 288. Pitman, London.

Straughan, B., Ewing, R.E., Jacobs, P.G. & Djomehri, M.J. (1987): Non-linear instability for a modified form of Burgers' equation. *Numer. Meth. Part. Diff. Equns.* **3**, 51–64.

Straughan, B. & Walker, D.W. (1996a): Anisotropic porous penetrative convection. *Proc. Roy. Soc. London A* **452**, 97–115.

Straughan, B. & Walker, D.W. (1996b): Two very accurate and efficient methods for computing eigenvalues and eigenfunctions in porous convection problems. *J. Computational Phys.* **127**, 128–141.

Straughan, B. & Walker, D.W. (1997): Multicomponent diffusion and penetrative convection. *Fluid Dyn. Res.* **19**, 77–89.

Stuart, J.T. (1988): Non-linear Euler partial differential equations: singularities in their solution. In *Proceedings of the Symposium in Honor of C.C. Lin.* Eds. D.J. Benney, F.H. Shu & Chi Yuan. World Scientific, Singapore.

Stuart, J.T. (1991): The Lagrangian picture of fluid motion and its implication for flow structures. *IMA J. Appl. Math.* **46**, 147–163.

Stuart, J.T. & Tabor, M. (1990): The Lagrangian picture of fluid motion. *Phil. Trans. Roy. Soc. London A* **333**, 263–271.

Suzuki, R. (1995): Blow-up of quasilinear degenerate parabolic equations with convection. *SIAM J. Math. Anal.* **26**, 77–97.

Tanveer, S. (1996): Asymptotic calculation of three-dimensional thin-film effects on unsteady Hele–Shaw fingering. *Phil. Trans. Roy. Soc. London A* **354**, 1065–1097.

Thess, A., Spirn, D. & Jüttner, B. (1995): Viscous flow at infinite Marangoni number. *Phys. Rev. Letters* **75**, 4614–4617.

Trefethen, L.N., Trefethen, A.E. & Reddy, S.C. (1992): Pseudospectra of the linear Navier–Stokes evolution operator and instability of plane Poiseuille and Couette flows: (preliminary report). Technical Report 92-1291. Dept. Computer Science, Cornell University.

Vulliet, L. (1995): Predicting large displacements of landslides. In *Numerical Models in Geomechanics.* Eds. G.N. Pande & S. Pietruszczak. Balkema, Rotterdam. pp. 527–532.

Vulliet, L. & Hutter, K. (1988a): Viscous-type sliding laws for landslides. *Can. Geotech. J.* **25**, 467–477.

Vulliet, L. & Hutter, K. (1988b): Continuum model for natural slopes in slow movement. *Géotechnique* **38**, 199–217.

Watson, E.B.B., Banks, W.H.H., Zaturska, M.B. & Drazin, P.G. (1990): On transition to chaos in two-dimensional channel flow symmetrically driven by accelerating walls. *J. Fluid Mech.* **212**, 451–485.

Weissler, F.B. (1985): An L^∞ blow up estimate for a non-linear heat equation. *Comm. Pure Appl. Math.* **38**, 292–295.

Whitham, G.B. (1974): *Linear and Non-linear Waves.* Wiley, New York.

Wu, C.C. & Roberts, P.H. (1994): A model of sonoluminescence. *Proc. Roy. Soc. London A* **445**, 323–349.

Wu, Y.H. (1995): Global existence and non-existence problem for a reaction-diffusion equation with a conserved first integral. *Proc. Roy. Soc. London A* **451**, 701–709.

Xie, W. (1991): A sharp pointwise bound for functions with L^2-Laplacians and zero boundary values on arbitrary three-dimensional domains. *Indiana Univ. Math. J.* **40**, 1185–1192.

Yuen, D.A. & Schubert, G. (1979): The role of shear heating in the dynamics of large ice masses. *J. Glaciology* **24**, 195–212.

Yuen, D.A., Saari, M.R. & Schubert, G. (1986): Explosive growth of shear-heating instabilities in the down-slope creep of ice sheets. *J. Glaciology* **32**, 314–320.

Subject Index

Springer
and the
environment

At Springer we firmly believe that an
international science publisher has a
special obligation to the environment,
and our corporate policies consistently
reflect this conviction.
We also expect our business partners –
paper mills, printers, packaging
manufacturers, etc. – to commit
themselves to using materials and
production processes that do not harm
the environment. The paper in this
book is made from low- or no-chlorine
pulp and is acid free, in conformance
with international standards for paper
permanency.